GREAT ESCAPES

THE TIMES
GREAT ESCAPES

BARBARA A. BOND

Published by Times Books
An imprint of HarperCollins Publishers
Westerhill Road
Bishopbriggs
Glasgow G64 2QT
www.harpercollins.co.uk

First published 2015

© HarperCollins Publishers 2015
Text © Dr. Barbara A. Bond 2015
Illustrations and Images © see pages 255–257

ISBN 978-0-00-814130-1

10 9 8 7 6 5 4 3 2 1

The Times® is a registered trademark of Times Newspapers Ltd

All rights reserved. No part of this publication may be reproduced, stored in a retrieval system, or transmitted, in any form or by any means, electronic, mechanical, photocopying, recording or otherwise without the prior permission in writing of the publisher and copyright owners.

The contents of this publication are believed correct at the time of printing. Nevertheless the publisher can accept no responsibility for errors or omissions, changes in the detail given or for any expense or loss thereby caused.

HarperCollins does not warrant that any website mentioned in this title will be provided uninterrupted, that any website will be error free, that defects will be corrected, or that the website or the server that makes it available are free of viruses or bugs. For full terms and conditions please refer to the site terms provided on the website.

A catalogue record for this book is available from the British Library

Printed in Hong Kong

If you would like to comment on any aspect of this book, please contact us at the above address or online:

timesatlas@harpercollins.co.uk
facebook.com/thetimesatlas
@TimesAtlas

CONTENTS

Acknowledgements		6
Introduction		7
Chapter 1	The Creation of MI9	13
Chapter 2	Background to the Mapping Programme	33
Chapter 3	The Map Production Programme	45
Chapter 4	Smuggling Maps and Other Escape Aids into the Camps	85
Chapter 5	Coded Correspondence with the Camps	115
Chapter 6	The Schaffhausen Salient and Airey Neave's Escape	147
Chapter 7	Escaping Through the Baltic Ports	173
Chapter 8	Copying Maps in the Camps	189
Chapter 9	MI9 and its Contribution to Military Mapping	197
Appendices		215
Appendices 1–9	Maps Known to have been Produced by MI9: A Carto-Bibliography	216
Appendix 10	Decoding a Hidden Message	242
Bibliography		247
Illustration and Photographic Credits		255
Index		257

ACKNOWLEDGEMENTS

I am indebted to many people for the realization of this long held ambition. To many friends and colleagues in Plymouth University especially Professor Wendy Purcell, formerly the Vice-Chancellor, for her inspiration, belief and confidence, Associate Professor Harry Bennett, Professor Mark Brayshay and Professor Kevin Jefferys for their outstanding scholarship, support and encouragement, Professor David McMullan for sharing his knowledge of cryptography, Tim Absalom and Jamie Quinn for their cartographic expertise, Andy Merrington and Lloyd Russell for photographs; to Peter Clark, formerly the Chief Map Research Officer in the Ministry of Defence, long-time professional colleague, mentor and friend, whose cartographic knowledge and expertise, and willingness to share them, remains undiminished; to Paul Hancock, Brian Garvan and Jim Caruth, former colleagues in the Mapping and Charting Establishment RE (now the Defence Geographic Centre); to the custodians of the many map collections, record repositories and museums I have visited and talked to in my pursuit of the pieces of the jigsaw puzzle, not least The National Archives, British Library, National Library of Scotland, Macclesfield Silk Museum, University of Glasgow, Second World War Experience Centre, Royal Air Force Museum, and the Intelligence Corps Museum; to John E. Bartholomew and the late John C. Bartholomew, Nicola Shelmerdine, daughter of the late Flight Lieutenant John H. Shelmerdine DFC, and Stephen Pryor, son of the late Commander John Pryor RN, for allowing access to family archives; to the late Professor M. R. D. Foot for sharing his experiences and knowledge.

My deepest thanks go to them all.

This book is dedicated to Roger, Abi and Ben, Adam and Johanna, for their love and support.

INTRODUCTION

'It is the intelligent use of geographical knowledge
that outwits the enemy and wins wars.'
(W. G. V. Balchin in *The Geographical Journal*, July 1987)

This book is the culmination of the author's personal fascination with maps and charts, and especially with military maps on silk. To be given the task, as a young researcher in the Ministry of Defence Map Library over thirty years ago, of identifying MI9's escape and evasion maps and creating an archival record set of them was a piece of serendipity. That small task ignited an interest which has never been extinguished. It was, however, only after retirement that the opportunity arose to commit to a detailed and more in-depth study of the subject: the reward was the discovery of a story of remarkable cartographic intrigue and ingenuity, and the opportunity to make this small contribution to the history of cartography in the twentieth century.

During the course of World War II, a complex and daring operation was launched by MI9, a newly formed branch of the British intelligence services, to help servicemen evade capture and, for those who were captured, to assist them in escaping from prisoner of war camps across Europe. Ingenious methods were devised to deliver escape and evasion aids to prisoners, and intricate codes were developed to communicate with the camps. In stories that often appear stranger than fiction, such materials proved critical and made many escapes possible. Maps were an integral part of this operation, with maps printed on silk and other fabrics commonly being secreted in innocent-looking items being sent to the camps, for example in playing cards, board games and gramophone records. The role of maps in this operation has often been overlooked and, because of strict instructions to service personnel at the time not to speak about the maps, the story has remained largely untold.

The principal aim of this book is, for the first time, to reconstruct, document and analyse the programme of escape and evasion mapping on which MI9 embarked. Such an exercise has never previously been attempted. The book charts the origins, scope, nature, character and impact of MI9's escape and evasion mapping programme in the period 1939–45. It traces the development of the mapping programme in the face of many challenges and describes the ways in which MI9 sought to overcome those challenges with the considerable assistance of both individuals and commercial companies. Through a number of examples, the extent to which the mapping programme was the key to the success of the whole of MI9's escape programme is assessed. The Appendices contain a detailed carto-bibliography, where all the individual maps are identified and described; production details are provided and location information on those surviving copies which have been identified is also provided.

The new intelligence branch was born in December 1939 and was charged with escape and evasion activities to support those who, it was anticipated, would either be shot down in enemy-held territory or would be captured. Its gestation had been lengthy. From the view held prior to World War I that there was something ignominious about capture, the military philosophy had evolved sufficiently in the inter-war period for escape activity to be regarded as a priority in the greater scheme of warfare. The new branch tried to tackle the aftermath of the disaster which befell the Expeditionary Force at Dunkirk in May 1940, but it took time and resources to mount the sort of organization which was needed. By then the nation was led by a new Prime Minister, Winston Churchill, who had had personal experience of escaping from captivity in the Boer War and was, undoubtedly, a natural ambassador for the changed philosophy. He had been a war correspondent for the *Morning Post* and was captured by the Boers in November 1899. He escaped on 12 December, and, travelling on foot and by train, he successfully made it across the border to Mozambique and freedom. In his account he noted that he was 'in the heart of the enemy's country' but lacked 'the compass and the map which might have guided me'.

The story of the mapping programme which became such an important part of MI9's escape programme has been difficult to piece together. No single, comprehensive record of the production programme has ever been found and the archival record set of the maps is only now (2015) being deposited in The National Archives by the Ministry of Defence. Copies of the maps have been found in many collections, both public and private, throughout the country. There is, however, very little mention of the maps in the published literature and some of the possibly relevant MI9 files in The National Archives are still closed. Reconstructing the story has proved to be like the reassembly of a jigsaw puzzle where some pieces are still missing, possibly lost for all time. Nevertheless, careful enquiry has yielded enough evidence to enable the narrative to be recovered. What emerges is a story of quite amazing inspiration and ingenuity in a country at war, fighting for the survival of its core democratic values and standards.

The intelligence world of the time also needs to be considered. MI9 was a new player in a world where others had already carved out for themselves a sizable role and niche. The part played by MI6, more commonly referred to as SIS (the Secret Intelligence Service), proved to be unhelpful and unsupportive to the newly created branch. The key staff recruited to MI9, especially Christopher Clayton Hutton, who was ultimately responsible for the mapping programme, had some of the skills they needed, but absolutely no cartographic awareness. The dark fog which often surrounded the world of security and espionage meant that key contacts were not made at critical times and there was often a lack of awareness of where the experience and expertise which MI9 so badly needed actually lay.

Producing the maps was one thing, getting them to the prisoners of war was quite another. Just how they were smuggled into the camps, how MI9 communicated with the camps, and how the whole escape programme took shape is part of the unfolding story. The art of smuggling items inside hollowed out containers was used extensively by MI9, who persuaded the manufacturers of board games and other leisure aids to assist them in their endeavours. Coded communication with the camps proved to be the vital link in the chain and deciphering a number of surviving coded letters provided the proof of the importance of that system of communication. The prisoners of war themselves also rose to the challenge. They had been trained that it was their duty to attempt to escape, not least by MI9 itself, which promoted the philosophy of 'escape-mindedness' through its Training School. Escape committees were set up in many of the camps, certainly in the oflags (prisoner of war camps for officers). The many hours of the potentially excruciating boredom of captivity were funnelled into escape activity. Escapes were managed as military operations in both the planning and execution. Men used their talents and, in many cases, their professional expertise, to copy maps, to produce compasses and clothing, and to forge papers and passes to aid the escapers on their journey to freedom. Lessons were learned from both the successes and failures, and key experiences were either brought back to the camps by the failed escapers or relayed back to the camps by the successful escapers, for use in future attempts.

The structure of the book reflects the story of the mapping programme as it unfolded. Chapter 1 shows that MI9 was essentially a creation of World War II and it reflected a markedly changed military attitude to capture, escape and evasion. It was staffed with people who had been carefully selected by the head of the fledgling organization. The skills and experiences which they each brought to the task are described and acknowledged, and the development of the organization itself is traced. Chapter 2 looks at the background behind the development of the mapping programme and the long history of military mapping on silk, which was most certainly not prompted simply by a twentieth-century war. Chapter 3 describes in detail the whole map production programme from the individual series, through the printing process to the sourcing of silk, and later, of man-made fibre. The covert nature of the programme and the compartmentalized way in which it was managed by MI9 resulted in their own unique, and arguably unnecessary, challenges. Having documented and analysed both the production programme and the maps themselves, the book continues in Chapter 4 by addressing the whole escape aids programme. The sheer ingenuity and originality of the smuggling programme which MI9 mounted in order to ensure that the maps reached their destinations is addressed. The whole of Chapter 5 is given over to a detailed discussion of the coded correspondence, augmented by the first deciphering of some of the original coded correspondence from a family archive.

A number of escapes were selected as examples to try to prove the value of the maps produced. The first of these, considered in Chapter 6, was based on one of the most successful escape routes which MI9 planned from occupied Europe to the safe haven of neutral Switzerland. Chapter 7 examines escapes via the Baltic ports to neutral Sweden, an even more successful route to freedom, and Chapter 8 studies the ways in which maps were copied in the camps and analyses two surviving maps apparently drawn in prisoner of war camps. Finally, Chapter 9 seeks to offer an objective assessment of the real success of the mapping programme in the light of the many obstacles and challenges which MI9 faced.

Without question, the maps produced by MI9 proved to be the key to successful escape: without them many, perhaps most, of the thousands of men who successfully escaped and made it back to these shores before the end of the war would have failed in their efforts.

The value of printing military maps on fabric has been long recognized. This map, printed on cloth, covers part of northwestern Georgia and adjacent Alabama to the west of Atlanta. It is annotated in blue pencil in the upper margin: 'Specimen of field maps used in Sherman's campaigns, 1864' (see pages 40–41).

Specimen of Field Maps used in Sherman's Campaigns, 1864.

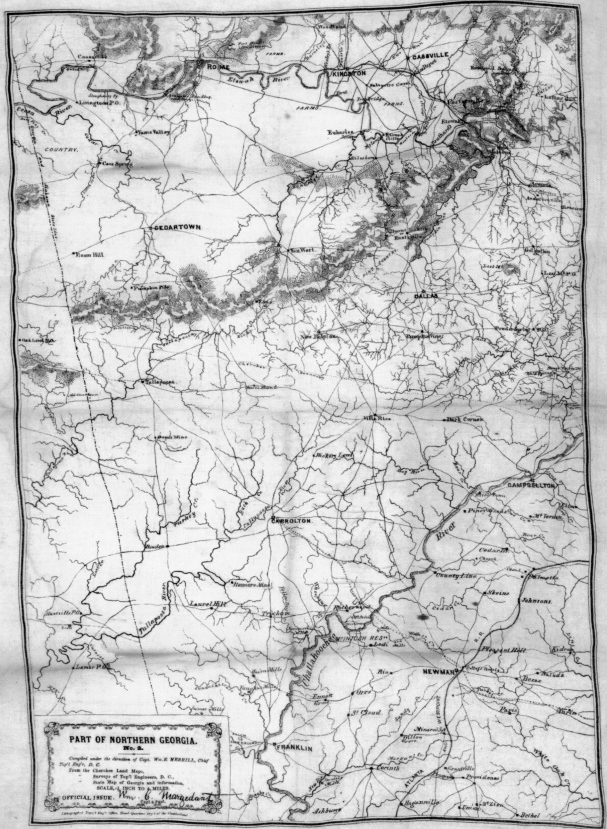

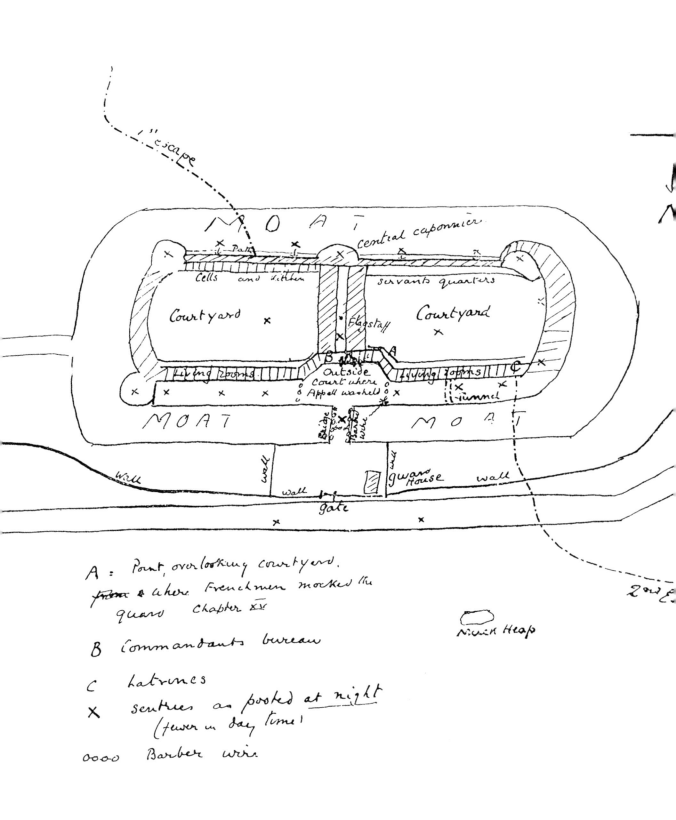

1

THE CREATION OF MI9

'Escaping and evading are ancient arts of war.'
(Field-Marshal Sir Gerald Templar, in the Foreword to *MI9: Escape and Evasion, 1939–1945* by M. R. D. Foot and J. M. Langley)

MI9 was created on 23 December 1939 as a new branch of British intelligence to provide escape and evasion support to captured servicemen and to airmen shot down over enemy-held territory through the course of World War II. Arguably, it was not soon enough, as, less than six months after its creation, thousands of British Service personnel found themselves captured on the beaches at Dunkirk. MI9 was established within the Directorate of Military Intelligence, which came into existence in 1939 when, with the Directorate of Military Operations, it superseded a previously combined Directorate of Military Operations and Intelligence. Five of the Military Intelligence sections, MI1 to MI5, continued their work within the new Directorate, dealing, as before, with organization, geographic, topographic, coded communication and security matters.

The creation of MI9 stemmed from the experience of many during World War I, when military philosophy about prisoners of war underwent a sea-change. From regarding capture and captivity in enemy hands as a somewhat ignominious, even shameful and disgraceful fate, the value that escaping prisoners of war might contribute to the success of the war effort gradually came to be recognized. Men who escaped or evaded capture and returned to Britain brought back vital intelligence and boosted the morale of the Armed Services and, not least, their own families. In addition, the considerable effort required to prevent escapes from the camps deflected the enemy's resources from front-line combat action.

In the late 1930s, as the prospect of war became increasingly likely, proposals for the creation of a section tasked to look after the interests of British prisoners of war came from many quarters, not least from Lieutenant Colonel (later Field-Marshal) Gerald Templar who had written to the Director of Military Intelligence in September 1939. A number of conferences with those who had been prisoners of war during World War I had also been arranged by MI1, seeking to benefit from their collective experiences. The actual proposal to the Joint Intelligence Committee to create such a branch came from Sir Campbell Stuart, who chaired a War Office Committee looking at the

Detail from a sketch map made by Johnny Evans (see page 22) showing his escape routes from Fort 9 Ingolstadt, during World War I. He brought his escape experiences with him when he joined MI9 in 1940.

Three British prisoner of war escapers who tunnelled out of Holzminden prisoner of war camp in Germany during World War I on 23 July 1918. That night twenty-nine men made good their escape, ten of whom made their way to the neutral Netherlands some 320 kilometres (200 miles) from the camp and eventually back to Britain. Left to right: Captain Caspar Kennard, Major Gray and Lieutenant Blair, all of the Royal Flying Corps.

Sir Campbell Stuart (1885–1972), who made the initial proposal for the creation of MI9 to the Joint Intelligence Committee in 1939.

coordination of political intelligence and military operations. There had clearly been some robust discussions, since Viscount Halifax, appointed Foreign Secretary in February 1938 by the Prime Minister, Neville Chamberlain, had indicated in a letter dated 5 December 1939 to Sir Campbell that his preference was for the section to be under Foreign Office control, with direct Treasury funding, presumably to ensure joint control and coordination with MI6, the Secret Intelligence Service (SIS). Notwithstanding this high-level opposition, the creation of MI9 went ahead in the War Office and it was made responsible to the Deputy Director for Military Intelligence, initially working closely with the Admiralty and the Air Ministry. With hindsight, the later animosity and conflict between MI9 and SIS (see Chapter 9) might well have had its roots in the initial difference of opinion as to where the newly formed section should sit in the governmental hierarchy.

MI9's objectives and methods were first outlined in the 'Conduct of Work No. 48', issued by the Directorate of Military Intelligence on 23 December 1939. In MI9's War Diaries (the regular record of daily, weekly or monthly activities undertaken by the War Office branches during the war), its objectives were more fully described as:

> i) To facilitate the escape of British prisoners of war, their repatriation to the United Kingdom (UK) and also to contain enemy manpower and resources in guarding the British prisoners of war and seeking to prevent their escape.

The Charter.

CONDUCT OF WORK No. 48.
M.I.9.

1. A new section of the Intelligence Directorate at the War Office has been formed. It will be called M.I.9. It will work in close connection with and act as agent for the Admiralty and Air Ministry.

2. The Section is responsible for:—

    (a) The preparation and *execution* of plans for facilitating the escape of British Prisoners of War of all three Services in Germany or elsewhere.

    (b) Arranging instruction in connection with above.

    (c) Making other advance provision, as considered necessary.

    (d) Collection and dissemination of information obtained from British Prisoners of War.

    (e) Advising on counter-escape measures for German Prisoners of War in Great Britain, if requested to do so.

3. M.I.9. will be accommodated in Room 424, Metropole Hotel.

(Sgd.) J. SPENCER.
Col. G.S.
for D.M.I.

23.12.39.

The original Conduct of Work No. 48 for MI9, produced by the Directorate of Military intelligence (DMI) and issued to MI5 and MI6, as it appears in Per Ardua Libertas, a photographic summary of MI9's work, produced by Christopher Clayton Hutton in 1942.

ii) To facilitate the return to the UK of those who evaded capture in enemy occupied territory.

iii) To collect and distribute information on escape and evasion, including research into, and the provision of, escape aids either prior to deployment or by covert despatch to prisoners of war.

iv) To instruct service personnel in escape and evasion techniques through preliminary training, the provision of lecturers and Bulletins and to train selected individuals in the use of coded communication through letters.

v) To maintain the morale of British prisoners of war by maintaining contact through correspondence and other means and to engage in the specific planning and execution of evasion and escape.

vi) To collect information from British prisoners of war through maintaining contact with them during captivity and after successful repatriation and disseminate the intelligence obtained to all three Services and appropriate Government Departments.

vii) To advise on counter-escape measures for German prisoners of war in Great Britain.

viii) To deny related information to the enemy.

The responsibilities included a mixture of operations, intelligence, transport and supply. The newly formed section was initially located in Room 424 of the Metropole Building (formerly the Metropole Hotel) in Northumberland Avenue, London, close to the War Office's Main Building.

NORMAN CROCKATT

The newly appointed Head of MI9 was Major, later promoted to Colonel and eventually to Brigadier, Norman Richard Crockatt (1894–1956), a retired infantry officer who had seen active service in World War I in the Royal Scots Guards. Crockatt had left the Army in 1927, worked in the City and was in his mid-forties at the outbreak of World War II.

Whilst he had been decorated in World War I (DSO, MC), he had never been captured and, therefore, had no experience of being a prisoner of war. He proved, however, to be an admirable choice to ensure the fledgling section made good progress in its infancy and throughout the war, being 'clear-headed, quick witted, a good organizer, a good judge of men, and no respecter of red

Brigadier Norman Crockatt in the MI9 headquarters at Wilton Park near Beaconsfield in 1944.

tape' (as recorded by M. R. D. Foot and J. M Langley in their definitive history *MI9: Escape and Evasion, 1939–1945*, hereafter referred to as Foot and Langley). These qualities were to stand him in good stead for the work he tackled in the next six years. He also recognized the importance of keeping his section small, concentrated in its activities and low profile among other intelligence sections, attributes which appeared to ensure that when the time came to expand its activities, it received little opposition from those competing for military priorities and budgets. Crockatt realized the value of having the experience of former prisoners of war, especially those who had successfully escaped, and appointed many with that experience to the small cadre of lecturing staff based in the Training School established by MI9.

The initial budget given to Crockatt to set up the entire section was £2,000. In present day terms, this equates to a sum of approximately £90,000. He embarked on a recruitment campaign and, by the end of July 1940, the complement of officers in the whole of the MI9 organization had risen to fifty. By that time, Crockatt was looking to move his organization out of London and by September 1940, he had selected Wilton Park, near Beaconsfield, as an appropriate location. After necessary refurbishment and the installation of telephones, most of the MI9 staff moved there on 14–18 October and occupied No. 20 Camp at Wilton Park.

ORGANIZATION

The section was initially organized into two parts: MI9a was responsible for matters relating to enemy prisoners of war and MI9b was responsible for British prisoners of war. The former subsequently became a separate department, MI19, to facilitate the handling and distribution of the intelligence information emanating from the two groups. On separation, the remaining MI9b was re-organized into separate sections and the staff complement was significantly increased:

> Section D was responsible for training, including the Training School which was established at the Highgate School in north London, from which the staff and pupils had been evacuated. It was subsequently designated the Intelligence School (IS9) in January 1942.
>
> Section W was responsible for the interrogation of returning escapers and evaders, including the initial preparation of the questionnaires which the interviewees were required to complete. The principal aim of the questionnaires was to identify information for use in the lectures and the *Bulletin*. The section was also responsible for the preparation and distribution of reports and for writing the daily, later to become monthly, War Diary entry.
>
> Section X was responsible for the planning and organization of escapes, including the selection, research, coordination and despatch of escape and evasion materials. Because of the small numbers of staff, the section was unable to spend much time on this activity until January 1942 when its establishment was boosted. At that point, they were also able to increase the volume of information to Section Y for transmission to the camps.
>
> Section Y was responsible for codes and secret communication with the camps. The development of letter codes as a means of communication with the camps was regarded as a priority from the start and the role which coded communication played in the escape programme is discussed in detail in Chapter 5.
>
> Section Z was responsible for the production and supply of escape tools, including all related experimental work.

It is primarily these last three sections whose activities largely, but not exclusively, form the focus of this book.

KEY STAFF

Christopher Clayton Hutton

Christopher William Clayton Hutton (1893–1965), known as 'Clutty' by all who worked with him, was appointed on 22 February 1940 as the Technical Officer to lead Section Z. He was the boffin, the inventor of gadgetry. His fascination for show business, particularly magicians, was apparently regarded as sufficient qualification for the post he was given as the escape aids expert in MI9. It was

Christopher Clayton Hutton, who led MI9's Section Z from 1940 to 1943, where he developed many ingenious escape aids and initiated the escape and evasion mapping programme.

his innate interest in escapology and illusion which was to prove the source of his imagination and ingenuity. Whilst working in his uncle's timber business in Park Saw Mills, Birmingham prior to World War I, he had challenged Harry Houdini to escape from a packing case constructed on the stage of the Birmingham Empire Theatre. Houdini escaped, for, unbeknown to Hutton, he had bribed the carpenter. In the inter-war years, he worked as a journalist and later in publicity for the film industry.

Hutton had served in the Yeomanry, the Yorkshire Regiment and as a pilot in the Royal Flying Corps during World War I. Realizing that another war with Germany was imminent, he tried to volunteer for the Royal Air Force and, subsequently, for the Army. When these approaches did not receive the encouragement he sought, he wrote a number of times to the War Office seeking an opening in an Intelligence Branch, an approach which eventually resulted in his appointment to the newly formed MI9 under Crockatt's leadership.

'Clutty' was regarded as both enthusiastic and original in his approach to the task to which he was appointed. He is variously described by his contemporaries as 'eccentric', 'a genius' and by Foot and Langley as 'wayward and original'. He wrote an account of his time in MI9, which ended during 1943 as a result of illness, under the title *Official Secret* (1960), but the publication of the book was not straightforward (see Chapter 9)

Crockatt very quickly recognized the value of having those who had experienced the reality of escape and was also conscious of the need to have representatives of all three Services in his organization. To this end, he appointed two Liaison Officers from the Royal Navy and the RAF.

Johnny Evans
From the RAF he appointed Squadron Leader A. J. Evans, MC (1889–1960). Johnny Evans, as he was always called, was an inspired choice. Appointed in January 1940, Evans very rapidly became a most valuable member of the MI9 team, becoming one of the star performers as a lecturer at the Training School. He had been an Intelligence Officer on the Western Front in World War I and had then been commissioned as a Major into the fledgling Royal Flying Corps (RFC). Shot down behind German lines over the Somme in 1916 and captured, he was eventually sent to the prisoner of war camp at Clausthal in the Harz Mountains.

For Evans, remaining in captivity was apparently never an option. Whether this in any way reflected his upbringing and public school education at

Winchester is unclear but it certainly did not reflect military training at the time. He escaped, only to be recaptured. On recapture he was sent to the infamous Fort 9 at Ingolstadt, north of Nuremburg, the World War I equivalent of Colditz in World War II. It was the camp to which all prisoners of war who had attempted to escape were sent and was located over 160 kilometres (100 miles) from the Swiss frontier. Evans described in considerable detail both his failed escape and his eventual successful escape in his best-selling book, *The Escaping Club*, published in August 1921, and reprinted five times by the end of that year. It took him and his companion almost three weeks to walk, largely at night, to the Swiss frontier which they crossed at Schaffhausen, west of Lake Constance.

Johnny Evans, from his classic book, The Escaping Club, *published in 1921.*

Evans described the attitude of the men in Fort 9 and the extent to which they spent their time in the all-consuming occupation of plotting to escape. It really was a veritable 'Escaping Club' where failed escapers were only too ready to share the knowledge of their experiences outside the camp with their

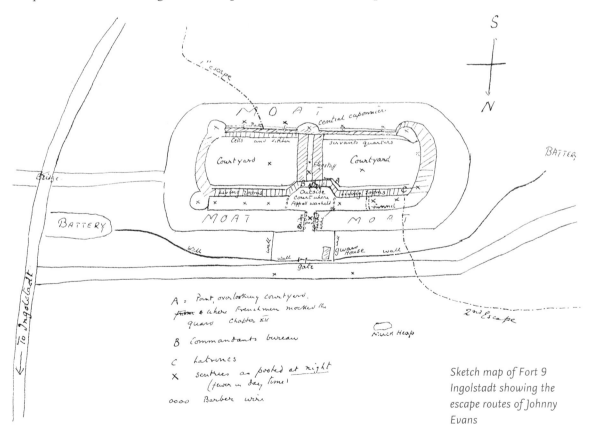

Sketch map of Fort 9 Ingolstadt showing the escape routes of Johnny Evans

fellow inmates. Evans described the receipt of clothes and food parcels from family and friends in which maps and compasses were also hidden. This was apparently accomplished by the prior personal arrangement of using a simple code in correspondence detailing the specific needs, maps, compasses, saws, civilian clothing and the like. Maps arrived in the camp secreted inside cakes baked by his mother or in bags of flour, and compasses inside bottles of prunes and jars of anchovy paste. The maps were copied in the camp so others could also use them and were then sewn into the lining of jackets.

It is fascinating to consider the extent to which Evans brought these experiences to bear in his work for MI9 and understandable that his book was dedicated:

> To MY MOTHER who, by encouragement and direct assistance, was largely responsible for my escape from Germany, I dedicate this book which was written at her request.

He went further, however, spending time prior to the outbreak of World War II visiting the Schaffhausen area of the German–Swiss border, across which he had made his own successful bid for liberty in World War I, photographing the border area and making copious notes. It cannot be coincidence that the MI9 *Bulletin* contained two large-scale maps of the Schaffhausen Salient, together with ground photographs of the local topography, showing distinctive landmark features, for example the stream and footpath where the German border guards patrolled. MI9's map production programme also included sheet Y, a large-scale map of the Schaffhausen Salient which carried very detailed notes of the topography and landscape features to help escaping prisoners of war (see Chapter 6 for the significance of this map in the MI9 programme and Appendix 1).

Jimmy Langley

Lieutenant Colonel James Maydon Langley (1916–83), called 'Jimmy' by friends and colleagues, was born in Wolverhampton, educated at Uppingham and Trinity Hall, Cambridge. As a young subaltern in the 2nd Battalion Coldstream Guards, he was badly injured in the head and arm and left behind at Dunkirk: on a stretcher as he would have taken up space which four fitter men could have occupied. He was taken prisoner, hospitalized in Lille and had his injured left arm amputated by medical staff. He subsequently escaped, in October 1940, by climbing through a hospital window and managed, with a still suppurating wound and with help from French families who befriended him, to reach

Jimmy Langley, who organized covert escape routes for MI9, and later co-authored the definitive history of MI9.

Marseilles from where he was repatriated, with help, through Spain to London. Langley successfully navigated by using the maps of the various French departments which appeared in every public telephone kiosk in France. The couple of times he travelled in the wrong direction resulted from the maps being oriented in a non-standard fashion, with East at the top. So Langley, too, knew the value of maps as an escape tool.

He arrived back in the UK in the spring of 1941 and initially joined SIS. He soon transferred to MI9 but remained on the payroll, and therefore technically under the command, of SIS. In practice he became the liaison point between the two organizations and was responsible for the work of the escape lines in northwest Europe. These were the covert routes along which the escapers and evaders travelled on their journey home, being helped by members of the local communities along the way. The network was established by MI9 working from London and also through legation staff and embassy attachés. If discovered by the Germans, as many of the French, Belgian and Dutch nationals were, they were executed or sent to the concentration camps. He remained in post for the duration of the war, and subsequently married Peggy van Lier, a young Belgian woman who had been a guide on the Comet Line (the organized escape route through western France and across the Pyrenees into Spain – see page 148). Over thirty years later, he co-authored with M. R. D. Foot *MI9: Escape and Evasion, 1939–1945*, which came to be regarded as the definitive history of MI9. He also wrote his own account of his early life, capture, hospitalization and subsequent escape in *Fight Another Day*, published in 1974.

Airey Neave

Of others in Crockatt's team, perhaps Airey Middleton Sheffield Neave (1916–79) is the most famous. As a young lieutenant, he was a troop commander in the 1st Searchlight Regiment of the Royal Artillery. He was wounded and captured in Calais Hospital in May 1940 as the Germans over-ran northern France. He was held in various oflags and, after a number of unsuccessful escape attempts, he was eventually imprisoned in Oflag IVC, the castle in Saxony more commonly known as Colditz. Working with his Dutch colleague, Toni Luteyn, Neave escaped from Colditz on 5 January 1942, the first British Officer to escape successfully from that infamous camp. After reaching Switzerland, Neave was repatriated through Marseilles and into Spain via the organized escape route known as the Pat Line (see page 148). After a short period of leave, he joined MI9 in May 1942 under the pseudonym (code-name) of Saturday. It is clear that his name had been on an MI9 list of targeted officers

Airey Neave was the first British officer to escape from Colditz.

who had been specifically helped to escape, and his experience was to prove invaluable. His escape and the extent to which it reflected the value of MI9's mapping programme are issues considered in detail in Chapter 6. His account, *They Have Their Exits* (1953), covered his experiences on the frontline, his capture and initial, unsuccessful attempts at escape followed by his escape from Colditz. Neave also wrote about MI9 in *Saturday at MI9* (1969), which is discussed in the Bibliography.

THE TRAINING SCHOOL AT HIGHGATE

Section D of MI9 was responsible for training and briefing the Intelligence Officers who attended courses at the Training School in Highgate and who then returned to their individual units to provide training to operational crews. Certainly in the early years, these training courses concentrated on the RAF whose crews were constantly overflying occupied Europe. The lecturers engaged were largely those who had personal experience of escape in World War I. Their remuneration was set at two guineas (£2. 2s. 0d.), the equivalent of around £82 today, for each lecture they delivered and they were also provided with travel and overnight hotel expenses when they travelled to deliver lectures at operational units. As early as January 1940, a conference was organized in Room 660 of the Metropole Building to hear a lecture delivered by Johnny Evans. By the end of February 1940, lecturers from the Training School had delivered their training lectures to seven Army Divisions, and five RAF Groups, and were undertaking a tour of the British Expeditionary Force (BEF).

The content of the general lecture given to officers and senior non-commissioned officers (NCOs) included emphasis on the undesirability of capture, instructions on evasion, conduct on capture and a demonstration of some of the aids to escape which were issued to units prior to deployment. The lecture emphasized that the job was to fight and avoid capture. If captured, it was their first and principal duty to escape at the earliest opportunity. Later on in the war, with the increasing numbers of prisoners of war and the increasing organization of Escape Committees in the camps, the lectures were updated to include mention of the Escape Committees, which were the responsibility of the Senior British Officer in each of the camps.

Those attending the training courses were told that money, maps, identity papers, provisions and many other escape aids would be made available through the Committees. The officers and NCOs who attended the lectures were then responsible for cascading the briefing down through the ranks, but they were, initially at least, specifically directed not to mention the aids to escape as they were only available for issue in limited numbers. It was recommended that they deliver the lecture as an informal talk, classified SECRET, and to audiences which should not exceed 200 at any one time. Later on, and certainly by early 1942, a supply of aids for demonstration purposes was provided to local commands.

There was also a classified TOP SECRET lecture on codes which was delivered under the title of 'Camp Conditions' to very limited audiences, never more than ten at a time, all of whom had been carefully selected. Those selected for this special briefing were required to practise the use of letter codes and their work was carefully checked before they were formally registered as authorized code users. Section Y was responsible for codes. The development of letter codes as a means of communication with the camps was also regarded as a priority from the start and the role which coded communication played in the escape programme developed apace. This aspect is discussed in detail in Chapter 5.

The staff in the Training School steadily compiled a training manual which became known as the *Bulletin*. The *Bulletin* served an important role as a tool in educating potential prisoners of war about possible escape routes and the nature of escape aids, including maps, which were being produced (see Chapter 4).

The pressures on the lecturing staff were considerable and continually increased as the war progressed. Initially both the Royal Navy and the Army had appeared uninterested in the training courses offered and, certainly in the first year or so of its existence, MI9 staff worked hard to stimulate interest and used many personal contacts to raise awareness of their work. They appeared to overcome some initial opposition from the Royal Navy and some Army commands, and by May 1944 the record shows that very significant numbers in all three services had been briefed: 110,000 in the Royal Navy and the Royal Marines, 346,000 in the Army and 290,000 in the Royal Air Force, and a total of 3,250 lectures had been delivered.

ESCAPE-MINDEDNESS

Escape-mindedness was the term which Crockatt coined to describe the philosophy which he sought to instil into the frontline forces which his staff regularly briefed and trained. Inculcating and fostering this philosophy was the primary aim of the training, and the rest of the MI9 team was working to ensure that the approach was supported in a very practical way. They stressed that, if

Stalag Luft III (Sagan), drawn by the artist Ley Kenyon, who was a prisoner in the camp. It shows the position of Tom, Dick, Harry and George tunnels. Harry was used in the 'Great Escape'.

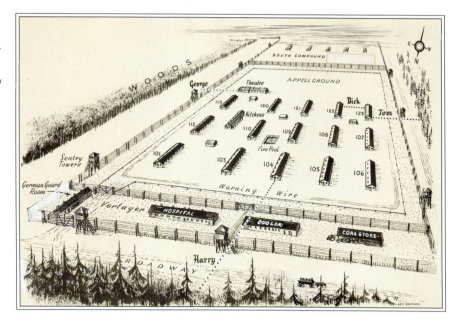

captured, it was an officer's duty to attempt to escape and, not only officers, it was a duty which extended to all ranks. Many years after the end of the war when Commander John Pryor RN came to write his memoirs of the years he spent as a prisoner of war during World War II, it is not surprising that he recalled that:

> escaping was the duty of a PoW but with the whole of NW Europe under German control and with no maps or compass it seemed a pretty hopeless task.

The briefings and training which MI9 provided alerted officers to every aspect of potential evasion and escape. The emphasis was on evading capture whenever possible or, if captured, to attempt to escape at the earliest opportunity and certainly before being imprisoned behind barbed wire in the many prisoner of war camps. It was standard practice for captured officers to be separated into oflags from the other ranks who were kept in stalags. Officers were, therefore, made responsible for ensuring that their men were appropriately briefed about what to do in captivity and the organization of Escape Committees became one of their principal priorities.

OVERLEAF:
A forged identity card made in the Holzminden camp in World War I and used by H. G. Durnford in his escape.

It is perhaps a reflection of the extent to which the philosophy permeated the camps that by the time the Allies were landing in occupied Europe and slowly advancing east, it was felt necessary to issue a 'stay-put' order to prisoners of war to ensure they did not get caught up in the frontline whilst trying to flee

THE CREATION OF MI9   27

Vor= und Zunamen: **Karl Stein**

Geburtstag: 4. Juni 1880

Geburtsort: Stralsund

Staatsangehörigkeit: Preussen

Grösse: 1,60.   Mund: gewöhnlich

Gestalt: untersetzt   Augen: braun

Kinn: gewöhnlich   Bart: Schnurrbart

Nase: gross   Haare: braun

Besondere Kennzeichen: —

Karl Stein
(Eigenhändige Unterschrift.)

Es wird hiermit bescheinigt, dass der Passinhaber vorstehende Unterschrift eigenhändig vollzogen hat.

STRALSUND, den 1. Mai 1918
DIE POLIZEI=SEKRETARIAT.
I.A.

Kozmick

captivity. The order was sent by MI9 on 18 February 1944 in a coded message: it directed that

> ON GERMAN SURRENDER OR COLLAPSE, ALL P/W ALL SERVICES INCLUDING DOMINION & COLONIAL & INDIAN MUST STAY PUT & AWAIT ORDERS

Many families also wrote to their sons in the camps strongly discouraging them from any escape attempts, as a result of the Stalag Luft III (Sagan) experience when fifty of the men who had taken part in the 'Great Escape' in March 1944 were executed on being recaptured.

The MI9 staff who subsequently wrote about their escapes, notably Neave and Langley, and even Evans who had escaped during World War I, all highlighted the importance of an escape philosophy. Neave described the way in which escapers had 'to think of imprisonment as a new phase of living, not as the end of life' and the extent to which the real purpose of the escaper was 'to overcome by every means the towering obstacles in his way'. It was a state of mind that MI9 encouraged.

It was understandable that some might prefer the relative safety of the camp rather than life on the run. Even for these men there were jobs to be done to support the escapes of others. It was strength of mind and purpose which was needed rather than just physical health and strength, a point epitomized by the escapes of Jimmy Langley, still suffering from a suppurating amputation wound, and Douglas Bader, restricted by his two artificial legs. Initiative, foresight and courage were needed and luck also came into it: as Evans stressed, 'however hard you try, however skilful you are, luck is an essential element in a successful escape', while David James noted in *A Prisoner's Progress* (1947) that:

> Luck is the most essential part in an escape . . . for every man out, there were at least ten better men who would have got clear but who did not have the good fortune they deserved.

Teamwork is the one competence which comes through all the stories and plans relating to escape. This almost certainly reflected the public school philosophy where your efforts were for school, house and team rather than for self. As an Old Wykehamist, Evans personified this approach and it is not surprising to learn that between the wars he captained the Kent county cricket team. To some extent it could be argued that MI9 was pushing at an open door in seeking to inculcate Crockatt's philosophy into a new generation of young men. Many of them had been educated at preparatory and public schools and apparently raised on a diet of escape classics of the last war. Some of them acknowledged this when they came to write their own accounts of their escape experience during World War II, as James recorded:

> In my prep-school days at Summer Fields, I had read all the escape classics of the last war – such books as *The Tunnellers of Holzminden*, *Within Four Walls*, *I Escape*, and *The Escapers' Club* [sic] – and as a proposition the business of escaping fascinated me.

It is clear from their post-war accounts that many escapers spent every waking moment of captivity plotting their escape. Some identified the very human traits which they believed could most aid them. Gullibility (of the captor) and audacity (of the escaper) were high on the list, as was luck. There was a psychology attached to escaping, as James recognized:

> I came to the conclusion that escaping was essentially a psychological problem, depending on the inobservance of mankind, coupled with a ready acceptance of the everyday at its face value.

The Germans were apparently well aware of this philosophy and the extent to which it sustained British prisoners of war and constrained their own resources in guarding those captured and seeking to prevent their escape. Once the Allies had landed in mainland Europe and started to advance east, they captured not simply German troops but also a number of key German documents amongst which was a document identified as GR-107.94. It must have made fascinating reading for MI9 as it revealed the extent to which the Germans were well aware of their work. It is a lengthy document and relates entirely to the escape methods employed by Allied Flying Personnel. It was dated 29 December 1944 and described the escape philosophy, the duty to escape, and the maps provided on silk and thin tissue. It goes so far as to list nine maps which they knew had been produced. Whilst it reflected the extent to which the Germans were aware of what they were up against, it also indicated that, if they were aware of only nine escape maps when MI9 had by that time produced over 200 individual items and over one and three quarter million copies, they had arguably only discovered the proverbial tip of the iceberg.

2
BACKGROUND TO THE MAPPING PROGRAMME

> 'For some time our engineers have been working on the problem of printing maps on cloth . . . the necessity of a durable material for maps was impressed on me a number of years ago . . . I was sent with my troop on an independent mission . . . about the second day, due to folding, use and the action of the elements, my map was almost illegible and I was travelling by a cavalryman's knowledge of the terrain.'
> (Lieutenant Colonel J. C. Pegram, Chief of the Geographic Section of the US War Department, in a letter dated 18 October 1927)

The story of the mapping programme has to be set in the climate of the times. The young men of the inter-war period, and especially the officers, most of whom had been educated in the British public school system, had been raised on a culture of escape stories from the Great War. They had read many of the books which had been written by the great escapers from World War I, people like Durnford, Evans and others. They had also been made more aware of the relevance of geography in their curriculum, of map reading and navigational skills. Their education had also sought to instil the standard British public school behaviour of team, country and King before self. They were avid readers of *Boy's Own Paper* and many had belonged to Baden-Powell's Boy Scout movement. Recognizing this, Christopher Clayton Hutton identified all the available literature, a total of fifty books (through a visit to the British Museum Reading Room) and purchased second-hand copies. He enlisted the support of the Headmaster at Rugby School, his alma mater, who allowed the sixth form to carry out a review of the books. The review was completed in four days, and led directly to Hutton's decision to make maps a priority, for it would appear difficult, if not impossible, to escape from enemy-occupied territory without a map. It was this simple fact which appeared to be the catalyst for Hutton's visit to the War Office Map Room. The staff there could not apparently help in meeting his initial request for a small-scale map of Germany.

The section responsible for operational maps and geographic matters, MI4, was by that time located in Cheltenham. It had moved from London in September 1939, apparently to make space to accommodate those branches whose presence in Whitehall was deemed to be essential and also to afford protection from possible air attacks to the sizable map collection which was also relocated to Cheltenham. Brigadier A. B. Clough, in his history of the

Detail from The Garrison Map, one of three silk maps excavated in 1973 from the Han Dynasty Tomb No. 3 in Mawangdui, Changsha, in Hunan Province, China (see pages 37–40).

military survey organizations during World War II, *Maps and Survey*, published in 1952, made it clear that the absence of MI4 from London 'had the serious effect of putting it out of daily touch with the General Staff at a critical period'. MI4 remained physically distanced from all War Office operations, intelligence and planning staff and also from the Air Ministry Map Section, which had been moved to Harrow. It is, therefore, likely that the War Office Map Room visited by Hutton was simply a small reference collection and not the main operational map collection of MI4 which would certainly have held the maps he sought. Hutton's lack of contact with the military map-makers is likely to have been to the longer term detriment of the escape and evasion mapping programme.

During a visit to the commercial mapping company, Geographia Limited, on London's Fleet Street, he discovered the existence of 'a famous Scottish firm' which proved to be John Bartholomew & Son Ltd of Edinburgh. This renowned cartographic company was established in 1826 by John Bartholomew, built on his and his father's experiences as apprentices to the Edinburgh engravers, Lizars, from the last years of the eighteenth century. By the late nineteenth century it had acquired a world-wide reputation for its maps. Hutton was also fortunate that the firm was headed at the time by John (known as Ian) Bartholomew who had had a distinguished military career in World War I, serving as an officer in the First Battalion, Gordon Highlanders, experiencing the worst of trench warfare and winning the Military Cross at Ypres in 1915.

Ian Bartholomew was only too ready to hand over copies of his company's maps, waiving all copyright and insisting 'it was a privilege to contribute to the war effort'. This was to prove the critical ingredient to MI9's wartime escape and evasion mapping programme. It was this collection of small-scale maps of Europe, the Middle East and Africa which provided the backbone of the escape and evasion mapping which MI9 subsequently produced. At the time, the company was not aware of its wartime involvement with MI9, a secret which Ian Bartholomew, the Managing Director, apparently never even mentioned to his sons.

MI9's War Diary entry for 31 March 1940 reflects just how quickly Hutton got to grips with the task he faced: the entry indicates that even by then, just three months into MI9's operations, available escape devices already included 'maps on fabrics and silk, maps concealed in games, pencils, articles of clothing'. Hutton had tried to find a paper which was thin, resistant to the elements and soundless when hidden inside Service uniforms, which was

John (known as Ian) Bartholomew, of the Edinburgh cartographic company John Bartholomew & Son Ltd, in the trenches near Ypres in 1915.

what he was planning to do. After talking to contacts in the trade, he became convinced that such a paper did not exist and so turned his attention to fabric, and to silk in particular.

RIGHT:
A printed Bartholomew map of France at 1:2M, used by MI9 as its 'Zones of France' map, but with southern England removed.

LEFT:
Detail from the same map.

OVERLEAF LEFT:
The Garrison Map covers the region between Mount Jiuyi and the Southern Ridges in Ningyuan in southern Hunan province, China. The map shows mountains, rivers and residential settlements, and in particular it indicates the locations of the garrisons, defence regions, military facilities and routes of nine army units. It is the oldest known map on silk, dating from the second century BC.

OVERLEAF RIGHT:
A reconstruction of the Garrison Map.

THE HISTORY OF MILITARY MAPS ON SILK

Hutton was almost certainly unaware that the efficacy of silk as a suitable medium for military maps had long been recognized. Indeed, the oldest surviving silk map in the world is a military map, known more commonly as the Garrison Map, excavated in 1973 from the Han Dynasty Tomb No. 3 in Mawangdui, Changsha, in Hunan Province, China. It was one of three silk maps found on the site, the others being a topographical map of the region and a city map. The Garrison Map has been dated to the middle of the second century BC. It was unearthed in twenty-eight fragments, moisture and pressure having taken their

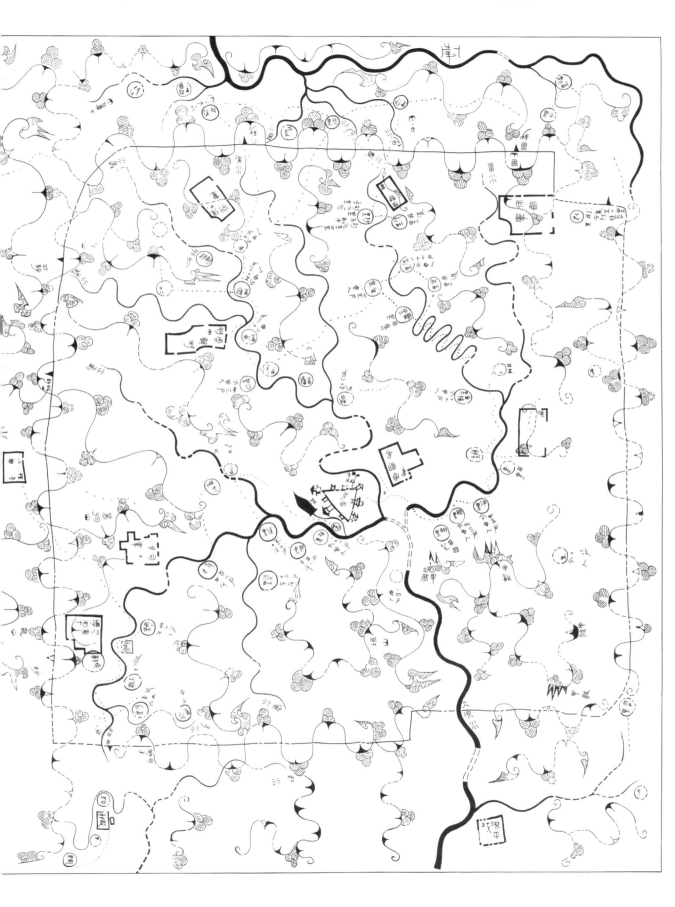

toll during 2,000 years of burial in a small box. The fragments were restored and then a reconstruction of the map was undertaken by Chinese scholars.

The map carries no indication of scale but, by comparing it with modern mapping, it is estimated to be in the range 1:80,000 to 1:100,000. It has been drawn on a rectangular piece of silk which measured 98 cm by 78 cm. The map was originally drawn in three colours, black, red and blue/green, using vegetable-based tints and is orientated and marked with south at the top and the left side marked east. Water features are shown in blue/green, with some background features and place names in black but the military content of the map is emphasized in red, showing the size and disposition of army units, command posts, city walls and watchtowers. Settlements are shown, together with the numbers of inhabitants. The boundary of the garrisoned area is marked and frontier beacons (observation outposts) are shown. Topographic detail is stylized, so that mountains are shown as wavy lines rather than by any attempt to represent their real form and shape. Roads are shown with distances between some settlements clearly marked, as are river crossing points: in modern military parlance, this is referred to as 'goings' or terrain analysis information and was a technique also utilized by MI9 in the production of some of their special area escape and evasion maps.

While the Chinese are understandably keen to stress the relevance of the three maps in terms of how they reflect their nation's achievements in surveying and mapping techniques in the wider context of historical cartography during the period of the Han Dynasty, the relevance here is that the Garrison Map was undoubtedly produced for military purposes and was drawn on silk. There are later examples of military mapping produced on silk in China. Seventeen hundred years after the Garrison Map was produced, the Garrison Outline Map of Shanxi was produced during the Ming Dynasty, although the later map was regarded by Chinese scholars as greatly inferior to the Han Dynasty Garrison Map, not least because the earlier map was drawn in colour and showed far more military detail than the Shanxi map.

In the USA during the Civil War, General Sherman was known to have had monochrome maps printed on cloth during the Atlanta campaign. The Library of Congress map collection contains many examples of Civil War maps printed on cloth, including the map illustrated here, showing 'Part of Northern Georgia', produced by the Topographical Engineer Office in Washington DC in 1864.

The Intelligence Division of the US War Department in Washington DC produced a map of Cuba in 1898 and one of China in 1900, both printed on

This map, printed on cloth, covers part of northwestern Georgia and adjacent Alabama to the west of Atlanta. It is annotated in blue pencil in the upper margin: 'Specimen of field maps used in Sherman's campaigns, 1864'. See page 11 for a larger image.

RIGHT:
Detail from this map.

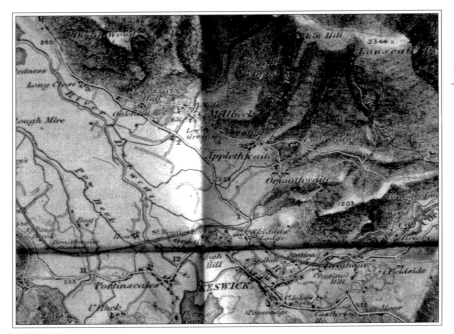

Detail from Ordnance Survey one inch map of the Lake District (Keswick), printed on silk and dating from 1891.

cloth. Details of these examples were contained in a letter dated 18 October 1927, written by Lieutenant Colonel J. C. Pegram, Chief of the Geographic Section of the Military Intelligence Division to Colonel R. H. Thomas, Director of Map Publications in the Survey of India in Calcutta, apparently in response to an enquiry from the latter. Interestingly Pegram highlighted the extent to which a medium more durable than paper was a necessity when the constant use, repeated folding and effect of the elements quickly rendered paper maps unusable in the field. He indicated that their engineers were currently addressing the problem of printing maps on cloth, were improving their techniques and getting good results, although he offered no technical details to support this statement. The challenge of printing on fabric was essentially that the cloth had to be held taut during the printing process so that the image would not be distorted.

The flexibility and durability of fabric, both silk and linen, as a medium for military maps had clearly long been recognized, a fact that had been reflected in the UK Government's report of the War Office Committee tasked in the closing decade of the nineteenth century to consider the precise form of the military map of the UK. The Committee made frequent mention in its Report, published in 1892, and throughout the minutes of evidence, to the superior durability of linen over paper as a material on which maps could be printed for use in the field. Almost certainly related to the Committee's work, although not acknowledged as such, the Ordnance Survey was simultaneously printing some of its one inch series of the Lake District on silk. Copies of two sheets, Ambleside and Keswick, are known to exist in a private collection and they are dated 1891. Ordnance Survey was still, at that time, staffed at senior levels by sapper officers from the Corps of Royal Engineers, so it is more than likely that they would have been called on by the War Office to do prototype printing experiments for military purposes. It is, however, notable that no mention of this work appears in the definitive history of the Ordnance Survey which concentrates rather on the work of the Dorington Committee which was taking place at the same time, charged with looking at the state of Ordnance Survey mapping after considerable public disquiet had been expressed.

With hindsight, it becomes very clear that the knowledge and capability to print maps on silk existed in the UK at the time of World War II and that the military had for centuries recognized the value of fabric maps, whether silk or linen, in terms of their durability and flexibility. Hutton's search and ultimate decision to produce maps on silk might have been made more promptly and with far less effort and cost had he spoken to the military map-makers. Certainly both the Directorate of Military Survey (D.Survey), the principal military mapping organization, and the Ordnance Survey doubtless had the expertise but, for unknown reasons, they were apparently never consulted by Hutton, who preferred rather to approach commercial printers, paper manufacturers and silk processors. This is rather surprising, bearing in mind the covert nature of MI9's activities and the secrecy which surrounded every aspect of their work, not least the mapping programme. It does, however, largely explain why MI9 paid little attention in their map production programme to the finer points of cartography and the standard techniques of identifying the maps they produced, as will be outlined in the next chapter.

3

THE MAP PRODUCTION PROGRAMME

'Geography is about maps ...'
(Edmund Clerihew Bentley, from *Biography for Beginners*, 1905)

The detail of MI9's mapping programme, which became such an important part of their escape programme, has been difficult to piece together. No single, comprehensive record of the production programme has ever been found and only now, in 2015, is a record set of the maps being deposited in The National Archives by the Ministry of Defence. Copies of the maps have been found in many collections, both public and private, throughout the country. There is, however, very little mention of the maps in the published literature and some of the possibly relevant MI9 files in The National Archives are still closed.

Reconstructing the story has proved to be like the reassembly of a jigsaw puzzle where some pieces are still missing, possibly lost for all time. The records that remain make it difficult to describe and record the extent of the programme and the challenges faced in ensuring the quality and utility of the maps for their intended users. It is likely that MI9 kept a card index of the individual maps in the programme in the same way as they are known to have kept a card index to maintain a record of other aspects of their work. Sadly, none of the card indexes appear to have survived and there is, therefore, no comprehensive record of MI9's escape and evasion map production programme available.

The programme has had to be pieced together from, sometimes, fragmentary information. The single most comprehensive record is undoubtedly the D.Survey war-time print record which was created, managed and kept up-to-date by Survey 2, that part of D.Survey responsible for the management of all operational map production programmes. This card index was originally found as an uncatalogued item in the India Office Library and Records, housed in the British Library. How it came to be held there, rather than in The National Archives where one might reasonably anticipate finding it, is an enigma, but it has now been catalogued by the British Library.

Detail from sheet 43B, showing part of the larger-scale inset of the German–Swiss border in the Schaffhausen Salient, (see pages 67–71 for more information on [Series 43]).

The second source is a typed list held in one of the War Office files. There is also a third source, a list deposited in the British Library by the archivist of John Waddington Ltd, the Leeds company that printed many of the maps, referred to as 'pictures' in their list and in their dealings with the Ministry of Supply. Intriguingly, the typed War Office list also refers to the maps as 'pictures', and it is, therefore, likely that it had its origins in the Waddington list.

A selection of escape and evasion maps produced by MI9.

The contents of the three sources are similar but by no means identical. Some of the differences can be explained by apparent human error (misreading of sheet numbers and typographic errors, for example). However, the War Office typed list contains other differences which are less understandable: for example, it states that twenty-nine sheets were produced in a series of maps of Norway, identified by the designation GSGS 4090, whereas the print card index indicates that the series consisted of thirty-three sheets, which proved to be correct since surviving copies of all thirty-three sheets have been identified.

While the print records were originally classified SECRET, some of the detail of the programme was apparently regarded as so sensitive that it was not declared openly even on that classified record. For example, it took some time to confirm, through seeing surviving copies, that the description of one map as 'D - - - - G' was a large-scale plan of the port of Danzig and that 'Dutch Girl', in four sheets, referred to Arnhem. Other entries on the record remain a mystery: for example, 'Double Eagle', although it might reasonably be conjectured that it was a map of Germany and Austria. Indeed, one of the earliest maps produced by MI9 was a small-scale map of Germany, Austria and adjacent frontier areas: it carried the sheet number A, probably indicating that it was the first sheet to be produced. No direct evidence has yet been unearthed, however, to confirm that this is the map which the record describes as 'Double Eagle'.

RIGHT:
Sheet A showing Germany and Austria at 1:2,000,000 scale, one of MI9's earliest silk maps, which used Bartholomew mapping. It may be the map referred to as 'Double Eagle' in the War Office print records.

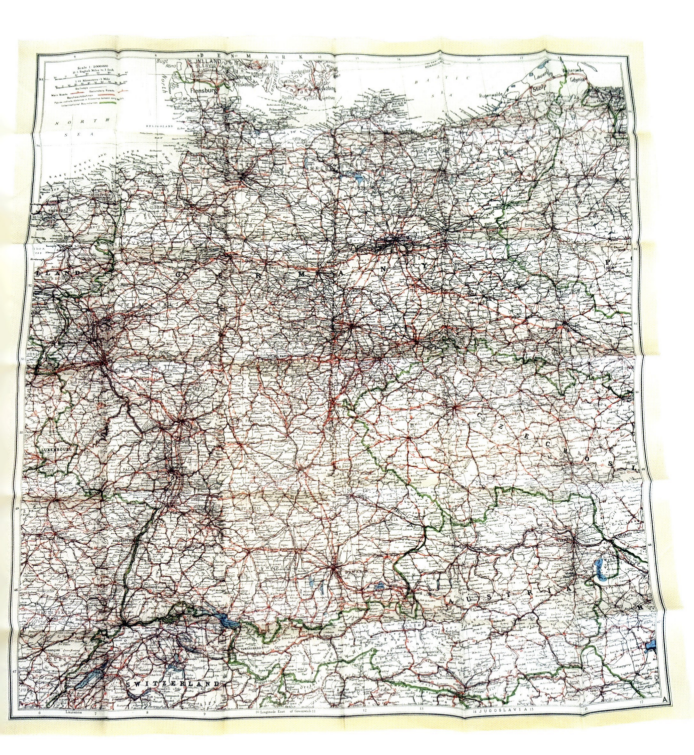

Where copies of individual sheets (either singly or in combination) survive, it has proved possible to identify the particular print medium, i.e. tissue (a very fine paper), silk or man-made fibre (MMF), almost always rayon but sometimes referred to as Bemberg silk. In some cases it has proved possible to be more precise. The Waddington list, for example, indicates that some of the maps were printed on Mulberry Leaf (ML) paper or Mulberry Leaf Substitute (MLS). Appendices 1–9 record the details of all the maps so far identified using these sources, including coverage, scale, dimensions, production details such as colour, print quantities and combinations, print medium, print dates and location of surviving copies.

MI9'S CARTOGRAPHIC INEXPERIENCE

MI9 may have been the initiators of the escape and evasion mapping programme but they were apparently not well versed in cartographic techniques, processes and procedures. Very many of the maps lack even basic identification. Many carry no title, series designation, date or edition number, and some carry no scale indicator. The proper referencing of military mapping is of fundamental importance in an operational scenario to ensure that everyone is using the same, and most up-to-date, version of the map. Even when cartographic referencing information is shown by MI9, it can prove to be

Sheet D of [Series 43] showing the title which identifies it as using 'New Frontier' and with the legend which identifies 'Former Frontiers' and 'Present Frontiers'.

48 GREAT ESCAPES

Bartholomew Sheet C.

very misleading. The most obvious examples of this are the sheets of GSGS Series 3982. The original operational map series designated with this GSGS (Geographical Section General Staff) number was the Europe Air series at a scale of 1:250,000 which existed prior to the outbreak of the War. These were reproduced as escape and evasion maps, largely on silk and tissue, at a reduced scale of 1:500,000. In reducing the scale, MI9 did not in any way alter the detail shown on the original map (even the series number) with the exception of the scale factor, so that on the resultant sheet, the font size of place and feature names is very small, although the detail is still legible.

The date on most of the sheets was in fact often the date of the original operational paper map and not the date of the escape and evasion map production. This is notably the case where compilation and imprint dates shown on the escape and evasion maps pre-date the start of the escape and evasion map production programme itself. This is also confirmed by the imprint numbers shown in the marginalia, often indicating print volumes well in excess of those produced as escape and evasion versions. The one exception to this was where boundaries are shown 'at 1943', for example in [Series 43]. Indeed, when D. Survey eventually assumed responsibility for escape and evasion map production in 1944, they proposed changes to the printing colours of boundaries and country names. In a letter dated 28 November 1944, MI9 came back strongly opposed to change, insisting that their policy of 'present frontiers in red and

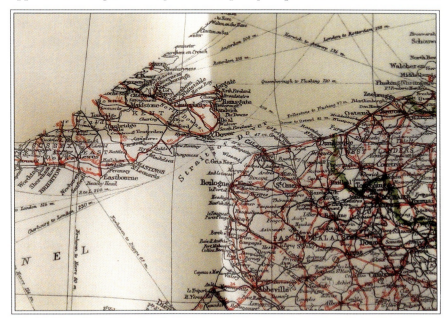

Detail from Bartholomew sheet C showing that MI9 stripped out most of the coverage of SE England in case copies fell into enemy hands.

THE MAP PRODUCTION PROGRAMME

Sheet A4 showing details of the port of Danzig.

pre-Munich frontiers in mauve' be adhered to, not least because it had always been specified as such in their training courses and lectures.

MI9's lack of knowledge of map production processes and procedures manifested itself in many other ways. It is a cardinal cartographic rule that different versions of the same map are identified differently, usually by a change in the edition number or, at the very least, in the production/print date. Both MI9 and the companies they initially used to print the maps were oblivious to such practices. As a result, the escape and evasion maps carried no edition numbers or production dates, and some maps apparently identified as being the same were in fact different. To give a few examples: there were at least two versions of sheet C with one version extending one degree longitudinally further east than the other version.

There were also at least three versions of the Danzig port plan. Whilst all three versions provided large-scale coverage of the port of Danzig, one carried the sheet number A4, whereas the second version carried no sheet number and the third carried the sheet number A3. The three versions varied also marginally in scale and in geographical extent. They also carried different intelligence annotations, the sheet marked A4 carrying far more intelligence information than the other two versions, in the form of annotations directing escapers, for example, where to find Swedish ships, where the arc lights were located and how far the beam of light extended. In the case of sheets J3 and J4 (covering Italy), the geographical areas of coverage of the two sheets were sometimes reversed

Detail showing the sheet identifier from sheet S3 of the Bartholomew series; sheet S2 is printed on the reverse so the identifier S3/S2 has also been included.

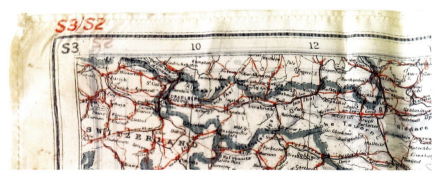

and the scale was varied, sometimes being produced at 1:1,378,000 and at other times reduced to 1:1,500,000 (see Appendix 1).

These are the principal variations identified to date: there may well be others yet to be discovered. It is important to describe them in detail since they are only really identifiable when surviving copies of the maps can be compared. The differences do, however, highlight the extent to which individual sheet identification of MI9's escape and evasion maps needs always to be treated with caution.

The solution to the challenges posed by MI9's lack of adherence to usual cartographic identification procedures has been to use the standard cartographic technique of showing in square brackets [ ] any information which does not appear on the printed maps but which helps to identify them. The series number and series title, for example [Series 43] and Series GSGS 3982 [Fabric], have been rendered in this form to aid identification.

Where the base map used for escape and evasion map production carried no sheet number, MI9 devised an arbitrary sheet numbering system. In the case of their early attempts based on the maps of John Bartholomew & Son Ltd of Edinburgh, the sheets carried an upper case Roman alphabet letter, often in conjunction with an Arabic number, for example C, H2, or S3.

MI9 caused themselves significant production problems when they decided (for unknown reasons) to cut and panel sheets to produce an escape and evasion map by piecing together up to nine sheet sections of an existing operational series, rather than simply reproducing the operational map sheets on their existing sheet lines. [Series 43] is a good example of this practice. It is clear that this escape and evasion series was produced by panelling together sheets or sections of sheets from the International Map of the World (IMW) series. There are examples of coverage diagrams in the surviving files of the composition of sheets created by this method. There are also indications that the practice caused considerable angst to the regular military map-makers when they eventually became involved in the escape and evasion production programme. On 3 December 1944, Lieutenant Colonel W. D. C. Wiggins of D.Survey wrote to MI9:

> Your proposed sheet lines do not (I have noticed this on previous layouts of yours) take into consideration existing map series sheet lines, printing sizes or fabric sizes . . . . Production is much simpler if we stick to graticule sheet lines as opposed to your, rather vague, rectangulars.

He might also have added that the practice must have greatly increased the production costs. The point appears to have been disregarded by MI9 who continued with arbitrary sheet lines and numbering systems, exemplified by [Series 43], [Series 44] and [Series FGS].

MAPS BASED ON BARTHOLOMEW MAPPING (AND OTHER MAPS WITH SIMILAR SHEET NUMBERS)

As Christopher Clayton Hutton indicated in his book, *Official Secret*, MI9 initially worked in isolation from the military map-makers and chose rather to approach commercial map publishing firms directly for help. As previously described, Hutton had contacted the firm of John Bartholomew & Son Ltd in Edinburgh at the suggestion of Geographia in London. It was Ian Bartholomew, the Managing Director, who gave Hutton his first lesson in map-making. Hutton himself indicated that 'thanks to the assiduities of the managing director and his staff . . . I learned all there was to know about maps'. Hutton was given copies of many of Bartholomew's own maps of Europe, Africa and the Middle East, which then formed the basis of MI9's initial escape and evasion map production programme. The waiving of all copyright charges for the duration of the war was a considerable financial gesture from Bartholomew since MI9 went on to produce in excess of 300,000 copies of the maps (details of the print runs are given in Appendix 1).

The maps are readily identifiable as using Bartholomew mapping since they are identical, in specification, colour and font style, to the company's maps of the time. They are generally small-scale (1:1,000,000 or smaller), produced in three colours (red, black and grey/green) and without elevation detail. A few of the maps carry confirmation of their source since they clearly show the Bartholomew job order number relating to the original paper map along the neat edge of the silk map. The alpha-numeric code A40 which appears in the northwest corner of some copies of sheet F was very much a

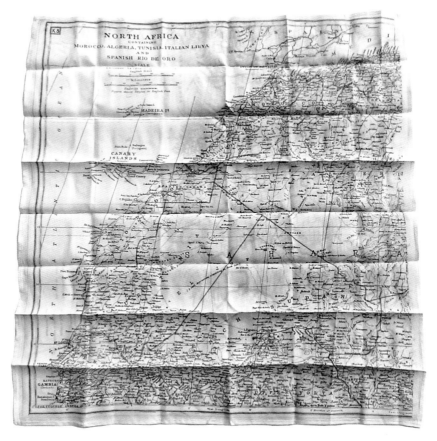

Sheet K3, printed on rayon, was based on Bartholomew mapping, primarily showing northwest Africa.

Bartholomew practice. The company introduced this code in the early part of the twentieth century, mostly on their half inch-scale mapping. The formula is a letter (either A, for January–June or B, for July–December) followed by a two-digit number representing the year of printing, so A40 indicates that the original paper version of this map was printed between January and June 1940.

Summary of Bartholomew series used by MI9

- ◊ 59 sheets identified with similar numbering
- ◊ 44 sheets based on Bartholomew maps
- ◊ 15 sheets use a similar numbering system but not based on Bartholomew mapping
- ◊ Coverage includes: Europe, Russia, Turkey, Middle East, North and East Africa, Scandinavia, South East Asia

- ◊ Scales: detailed maps 1:16,000 to 1:600,000 and regional maps 1:1,000,000 to 1:6,000,000

- ◊ Print dates identified: 7 January 1942 to 9 August 1943

- ◊ Printed on: tissue, silk, paper, man-made fibre (MMF), rag lithographic paper, bank paper

- ◊ Bartholomew-based sheets printed largely in three colours: black, red, grey/green

- ◊ Copies printed: 348,570

For full details of the maps, see Appendix 1.

The existence of a direct link between MI9 and Bartholomew is shown in a company memorandum dated 3 August 1940 from Ian Bartholomew to the company's London office, on the subject of 'Captain Clayton Hutton' indicating that a letter had been received from him thanking them for the prompt attention in sending the plates MI9 had requested. Hutton's original letter is not in the file. The memorandum indicates that Bartholomew not only handed over printed paper copies of the maps but also provided printing plates for their on-going reproduction. Further confirmation is contained in the first version of Hutton's memoirs to be published (under a pseudonym) by the inclusion of black and white photographs of two silk maps which can be readily identified as sheets A and C in the MI9 inventory. The link between MI9 and Bartholomew is also confirmed by the existence in a contemporary Air Ministry file of a printed copy of a map which was identifiable as sheet A/Germany carrying a clear imprint of the Bartholomew company.

MI9 added to the maps what passed for a rather crude sheet identification system in the form of an upper case alphabet letter, often in conjunction with an Arabic number (for example C, H2, K3). However, even this practice was not consistent as the same numbering system was also applied to some sheets which were clearly not based on the Bartholomew small-scale maps. The large-scale sheet of the port of Danzig (A4) at approximately 1:16,000, the large-scale map of Schaffhausen (A6) and the medium scale sheets of Italy (J5 and J6 at 1:275,000, J7 and J8 at 1:110,000) are six such examples. Sheet A4 Danzig appears to be an amalgam of the detail from a British Admiralty chart with additional ground intelligence added in the form of intelligence annotations. (This particular map will be discussed in more detail in Chapter 7.) Sheet A6 of Schaffhausen is based on large-scale, native German and Swiss topographic maps of the border area and sheets J5, J6, J7 and J8 appear to be based on large-

Due to the nature of the fabrics used, the ink often bled completely through to the reverse of the sheet, as shown here with Bartholomew sheet A. Many of the maps were still printed on both sides however, making it very difficult for the escapers to decipher the detail.

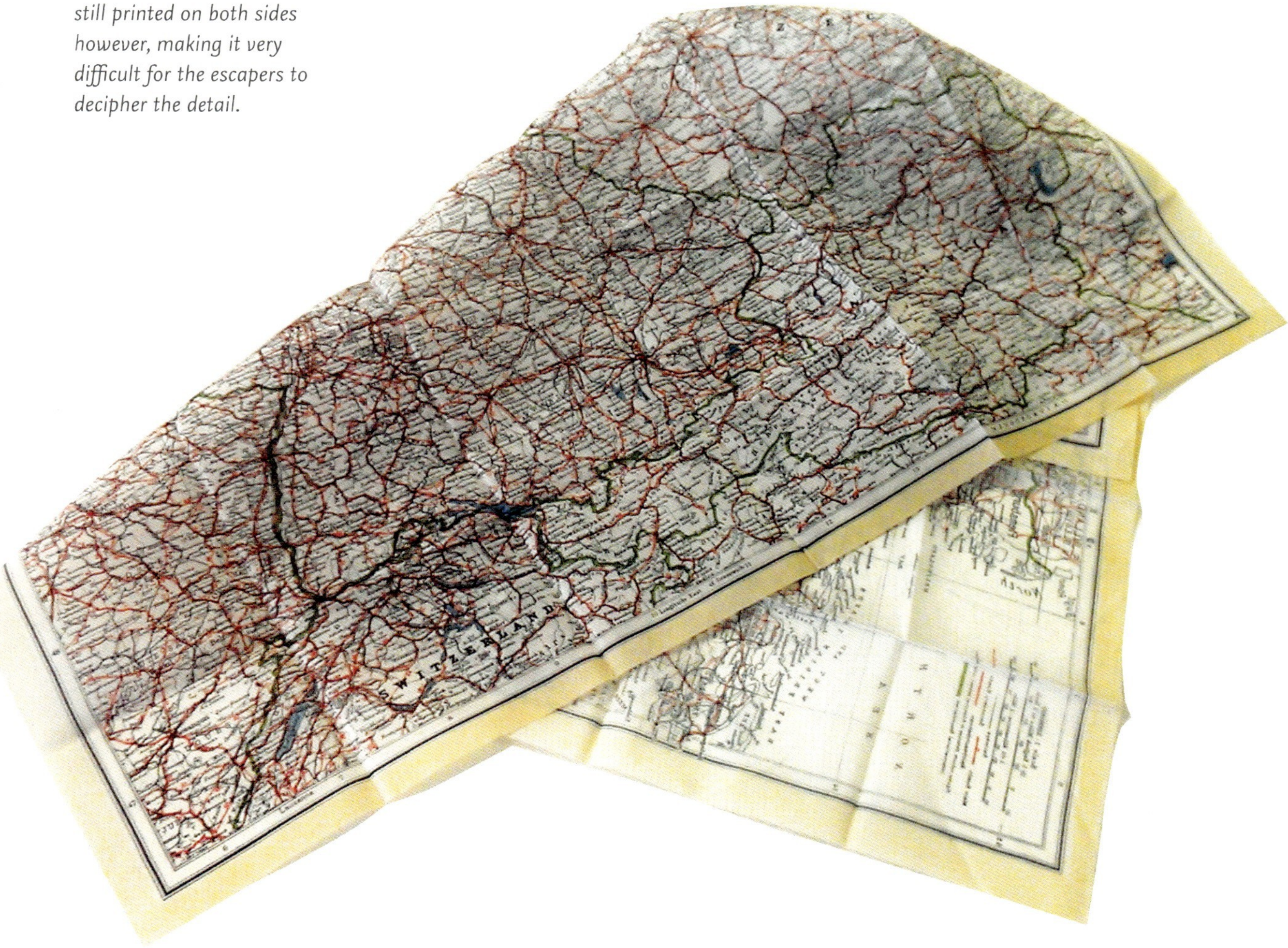

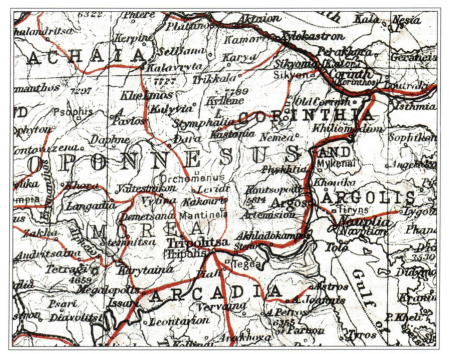

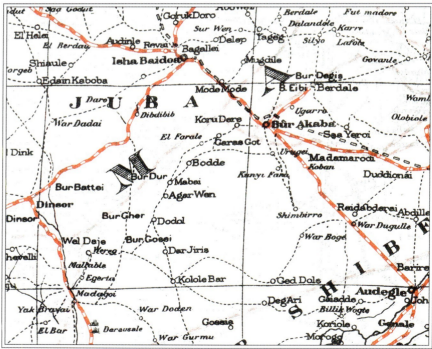

Details from a selection of Bartholomew maps showing their symbology, level of detail, density of place names and scale information. They also show the problems for the user caused by show-through from the map printed on the other side of the fabric.

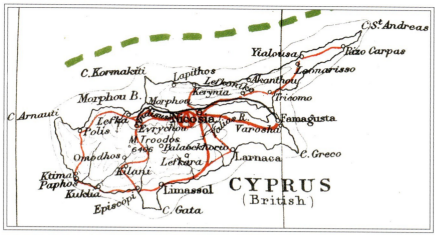

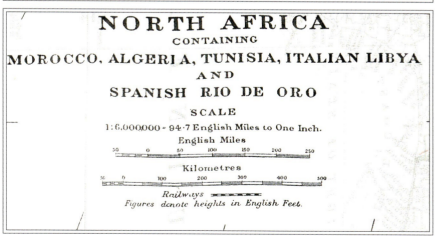

scale, native Italian topographic maps. For the purposes of this study, however, they have been identified as one series based on the similarity of the sheet numbering system. Certainly it appears that the first small-scale map of the area was identified with a single alphabet letter, such as A, and that any subsequent map produced which was located in the same geographic area was numbered A1, A2, etc. in sequence. To prove that this was, indeed, the approach adopted would assume that, since sheets J5 to J8 all provided large-scale coverage of the Italian area, sheet J should be a small-scale sheet of Italy. This cannot, however, be proved conclusively, since no copy of sheet J has yet been found and no mention of it appears in the record.

A further map included in this series (but which had no identifying letter) was a General Map of Ireland, produced in two sheets, printed back-to-back. This was almost certainly the map mentioned in the records as 'Shamrock', 500 copies being printed on fabric on 16 February 1942. There was apparently some discussion about the position of internees in the Republic of Ireland since, under the Geneva Convention, internees were held until the end of the conflict. On 5 March 1943, in a MOST SECRET internal minute from MI9 to the Director of Intelligence, a discussion between Sir John Maffey, the British representative in Eire, and the Irish Government was reported. It concerned the possibility of faking the escape of British prisoners of war interned in the Curragh. The deal fell through, apparently because the Irish Government wanted fighter aircraft in exchange. Whether the internees might be encouraged to escape anyway was a moot point since there was the distinct possibility of political embarrassment. The fact that the General Map of Ireland was produced at all appears to bear testimony to the fact that MI9 prepared for the possibility of escape and it was certainly recorded in the War Diary that nine RAF officers escaped from the Curragh Camp in Eire on 25 June 1941. Three were recaptured and six reached England. MI9 apparently wrote a special SECRET report on the escape, but this has not been identified.

Those sheets where surviving copies have been located in British record repositories and the various states in which the sheets were produced, either singly or in combination, are detailed in Appendix 1. A total of fifty-nine sheets have been identified in this initial group of the escape and evasion maps, of which over forty are clearly based on Bartholomew maps. Some of these carry the original Bartholomew job order number which has allowed them to be compared directly with the lithographic paper copy held in the company's print archive now housed in the National Library of Scotland's map collection in Edinburgh. Sixteen sheets which are believed to have been produced but for which no copies have been identified to date in British record repositories are listed in Appendix 2. This group is either noted on the print records, shown on adjacent sheets diagrams, represents gaps in the assumed consecutive numbering sequence or, in one case, has been spotted by a colleague at a map fair but no details were recorded.

Summary of fabric maps presumed to be based on Bartholomew originals

- ◊ These are maps which are assumed to have been produced based on Bartholomew maps but for which neither copies nor related records have ever been found
- ◊ 16 sheets identified
- ◊ Coverage includes Europe, Scandinavia and parts of Africa

For full details of the maps, see Appendix 2.

By autumn 1942, there was an increasing awareness that the small-scale maps being issued to aircrew were regarded by them as too small-scale. The need for better maps at larger scales gained momentum and 'a new series gained approval'. It is not clear what this new series was, but it is likely to have been the fabric versions of series GSGS 3982 (see the following section), as they started to be produced at that time, their earliest print dates being the late summer and autumn of 1942. At this stage valuable information about the Swiss frontier was also being incorporated into a map of the frontier area produced at 1:100,000 scale.

Double-sided Bartholomew sheet K3/K4. The nature of the fabrics used for MI9 maps allowed them to be folded into very small sizes.

EUROPE AIR 1:500,000 GSGS 3982 [FABRIC]

Despite the evidence indicating that MI9 engaged initially with commercial companies to develop and progress the escape and evasion mapping programme, there was clearly contact with the operational map-makers at some stage since any map series carrying a GSGS series number is immediately identifiable as an operational series. This group of escape and evasion maps was produced by reducing the original operational Europe Air series from 1:250,000 scale to 1:500,000 scale. This reduction could only have been done from the original reproduction materials which would have been held by D.Survey since, apart from their scale and print medium, they are identical in all respects to the original lithographic paper maps.

Summary of Europe Air 1:500,000 GSGS 3982 [Fabric]

- ◊ 74 sheets identified
- ◊ 4 additional irregular sheets, based on Arnhem ('Dutch Girl'); Sections 1, 2, 3, 4
- ◊ Coverage: Europe (see cover diagram opposite)
- ◊ Scale: 1:500,000; sheet N33/9 at 1:375,000; Arnhem maps at 1:420,000
- ◊ Print dates identified: June 1942 to August 1944
- ◊ Printed on: fabric, tissue, paper, mulberry leaf substitute (MLS), mulberry leaf (ML) tissue
- ◊ Printed in six colours
- ◊ Copies printed: 35,100 plus 8,221 Arnhem maps

For full details of the maps, see Appendix 3.

As a result of the scale reduction, the escape and evasion maps were one quarter the size of the original operational lithographic maps printed on paper. These maps were variously referred to as 'miniatures' or 'handkerchief' maps (while no confirmation of this description has been found in the records, it carries the ring of authenticity because of the small size of the maps and their reduced scale). The sheets were printed in six colours and individual sheet coverage is two degrees of longitude (°E) and one degree of latitude (°N). All geographical values of the extent and coverage of individual sheets have been derived from an index of the original series acquired from the Defence

Cover diagram of the Europe Air 1:500,000 GSGS 3982 [Fabric].

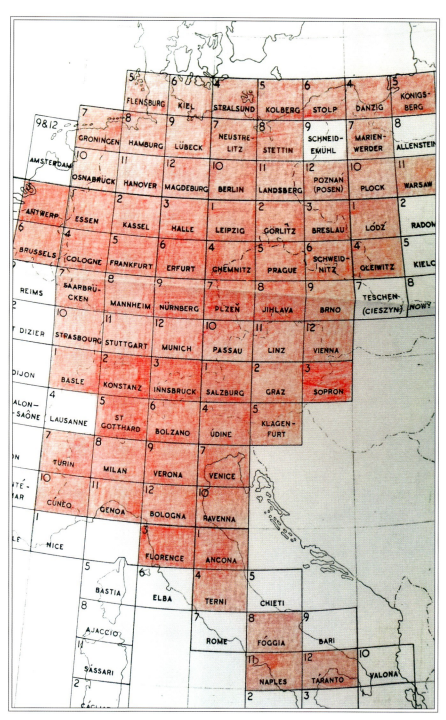

Turin, Europe Air 1:500,000 GSGS 3982 [Fabric], sheet L32/7.

LEFT:
Detail from this sheet, clearly showing the borders of Switzerland, France and Italy.

Geographic Centre (DGC). This index has also been used as the basis of the cover diagram shown on page 61.

Escape and evasion versions of some seventy-four sheets were produced. Over 35,000 copies were printed between 10 July 1942, when production apparently commenced, and an unrecorded day in July/August 1944, when the decision was taken to cease production. One can therefore conjecture that MI9's contact with D.Survey, either directly or using the Ministry of Supply as an intermediary, had started at least by mid-1942. Despite the significant number of copies produced, relatively few surviving sheets have been discovered. This may reflect the fact that many of the print runs were very small, 45 and 100 being common.

The sheets in this series were produced sometimes singly and sometimes in combination, although it has not proved possible to identify from the records the various combinations which were produced. They were produced on fabric or paper, the latter being variously described in the print record as 'RL', an acknowledged abbreviation for rag lithographic paper, 'thin BANK paper', ML (Mulberry Leaf) or MLS (Mulberry Leaf Substitute). One interesting group of a block of four of these sheets was produced under the codeword 'Dutch Girl'. They are centred on Arnhem and are of non-standard geographical extent, location and scale, being 1 inch to 6.56 miles or approximately 1:420,000; as a result they are referred to in the record as Sections 1, 2, 3 and 4. The record shows that some 4,400 copies of these four sheets were produced in June 1942 and were described by Hutton in earlier correspondence with Waddington as 'a very, very urgent requirement'. The purpose of producing these sheets at that stage is not clear, although it might have related to the combined operations which took place at that time and are known to have involved Section X of MI9 and the Special Operations Executive (SOE). In all probability, the subsequent printing of the same sheets, which took place in the period prior to August 1944, was in timely support of the Battle of Arnhem, otherwise referred to as Operation Market Garden, which took place in September 1944.

NORWAY 1:100,000 GSGS 4090 [FABRIC]

Once again MI9 must have acquired access to the reproduction material of another operational map series produced by D.Survey. Thirty-three sheets of this pre-war GSGS series of Norway at 1:100,000 scale were printed on silk, apparently for escape and evasion purposes, in 1942. The sheets are located in a block to the north of Oslo and adjacent to the Swedish border. Thirty-one sheets are monochrome, based on the 1940 state of the original GSGS 4090 sheets, while two sheets (26B and 26D) are printed in four colours (black, red, brown, blue) and are based on the 1942 revision of the GSGS 4090 sheets. The original lithographic paper maps were based directly on original native Norwegian maps, some of which dated back to the early years of the twentieth century. The original Norwegian maps were based on the Oslo meridian. The GSGS series maps, therefore, carry a conversion note to the effect that the Oslo meridian is 10° 43' 23" E of Greenwich.

Rjukan, Norway 1:100,000 GSGS 4090 [Fabric], sheet E35 West.

RIGHT:
Detail from this sheet.

Summary of Norway 1:100,000 GSGS 4090 [Fabric]

◊ 33 sheets identified

◊ All sheets single-sided

◊ Coverage: Norway north of Oslo and adjacent to Swedish border

◊ Scale: 1:100,000

◊ Print dates identified: 1942

◊ Printed on: silk

◊ 31 sheets printed in monochrome; two sheets in black, red, brown and blue

◊ Copies printed: not known

For full details of the maps, see Appendix 4.

It is believed that the purpose of this particular group of escape and evasion maps was specifically to help Allied airmen who were shot down and bailed out to evade capture and reach neutral Sweden. It is also known that crews flying damaged aircraft which they knew were unlikely to return home, were briefed to try, if they could, to put them down in southern Sweden. It is possible, therefore, that the maps were issued during pre-flight briefings only to those members of RAF crews destined to overfly Norway or adjacent parts of Denmark and Germany. It is, however, also possible that the maps related to combined operations discussed by SOE and MI9's Section X, scheduled for summer and autumn 1942. Section X had only been established in January 1942, specifically for planning escapes but also apparently for providing support for other intelligence-mounted operations. The operation mounted in Norway was

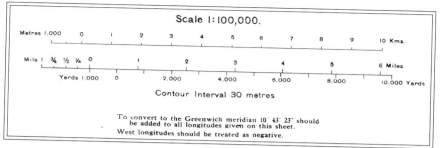

Extract from the marginalia of sheet 19D, showing the scale bars and the instructions on how to convert geographical coordinates from the Oslo meridian to the Greenwich meridian.

described as 'mysterious' in SOE terms, referred to only by its code name (Grouse) and scheduled to take place sometime after the end of September 1942. It related to the heavy water plant, the Norsk Hydro facility at Rjukan, on which SOE had mounted a huge intelligence-gathering operation as a precursor to blowing it up. It was known that this plant was producing heavy water which the Germans were planning to use in the manufacture of atomic weapons: this was undoubtedly the reason that it was such a strategic sabotage target for SOE. It cannot be coincidence that the coverage provided by this group of silk maps included sheet E35 West, the Rjukan sheet, providing detailed coverage of the plant's geographic location and adjacent area. The site had been chosen by the Norwegians during their pre-war construction of the plant because of its isolation 'between Vermok and Rjukan, in the precipice and glacier-bound wilderness of Hardanger Vidda', as described by Leo Marks in his book, *Between Silk and Cyanide*. Copies of all sheets still survive. A detailed list of those sheets in this series known to have been produced as escape and evasion maps on silk is given in Appendix 4.

The hydro-electric power plant at Rjukan, where German scientists attempted to produce the 'heavy water' necessary for the manufacture of atomic weaponry. The Germans were experimenting with the process as early as 1940, but the project was sabotaged by Norwegian and British forces.

Sheet 43A, showing northern France and parts of Belgium and Holland. The main map was printed at 1:1,000,000 with insets of the French–Spanish border at a larger scale.

TOP OVERLEAF:
Detail from sheet 43B, showing part of the large-scale inset at 1:300,000 of the German–Swiss border.

BOTTOM OVERLEAF:
Detail from sheet 43B showing specific information about the French–Spanish border in the legend.

[SERIES 43]

The use of standard operational mapping continued into 1943. This particular escape and evasion series had no title or individual sheet names. All sheets had the prefix 43 which, it is believed, refers to 1943, the year in which production commenced, and so this escape and evasion series has been identified as [Series 43]. A detailed list of the sheets, which are all small-scale (1:1,000,000) with some large-scale insets, is provided in Appendix 5. There are ten basic sheets, produced largely in combinations, with only one sheet apparently printed single-sided. The coverage extends from 10°W to 26°E (Portugal to Turkey) and from 40°N to 52°N (central Spain to Holland).

Summary of [Series 43]

◊   10 sheets identified

◊   Double-sided combinations produced: 8 identified

THE MAP PRODUCTION PROGRAMME   67

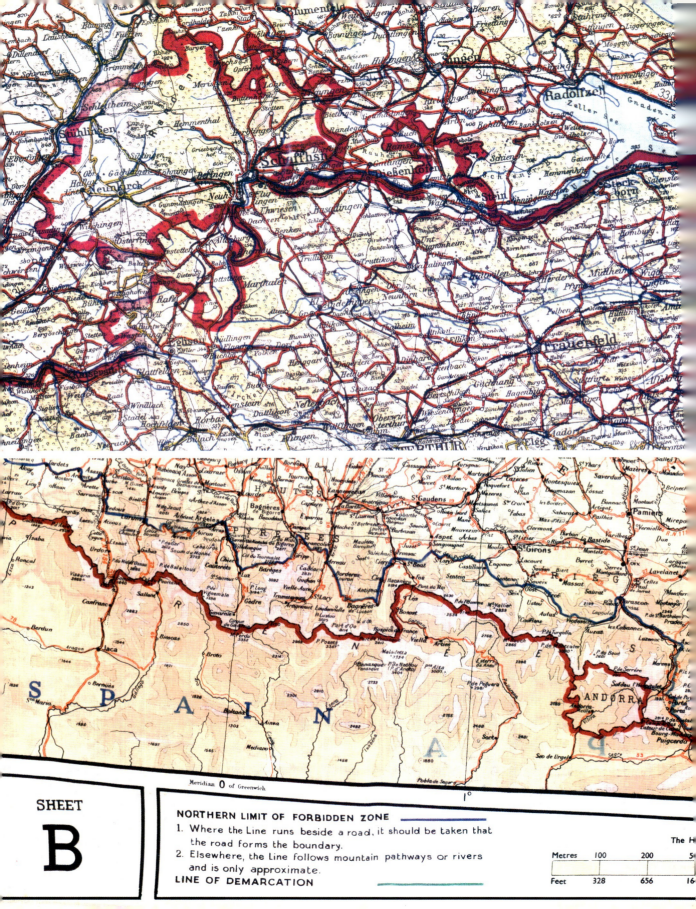

SHEET **B**

NORTHERN LIMIT OF FORBIDDEN ZONE ─────
1. Where the Line runs beside a road, it should be taken that the road forms the boundary.
2. Elsewhere, the Line follows mountain pathways or rivers and is only approximate.

LINE OF DEMARCATION ─────

Metres	100	200
Feet	328	656

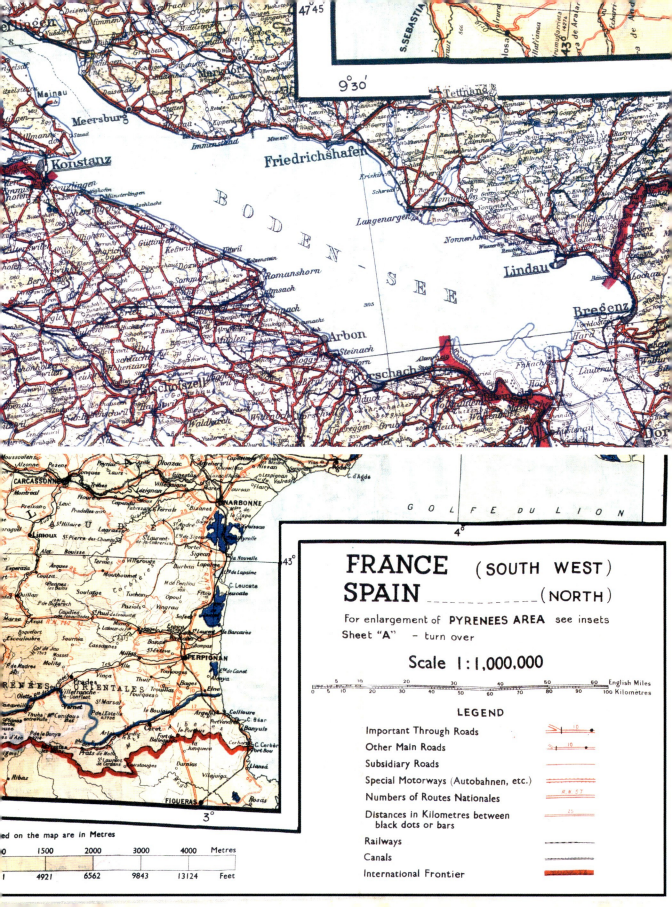

- ◊ Coverage: Europe, with insets of the Pyrenees, German–Swiss frontier, Belgian–German frontier

- ◊ Scale: 1:1,000,000 with insets 1:250,000 to 1:500,000

- ◊ Print dates identified: 1943 and 1944

- ◊ Printed on: man-made fibre (MMF)

- ◊ Multi-colour printing

- ◊ Copies printed: 1,159,500

For full details of the maps, see Appendix 5.

It is notable that the records indicate that over one million copies of these sheets were produced. It is possible that, as with [Series 44], they were produced for operational use in the field as well as for escape and evasion purposes (see the following section for details of the proposed operational use of escape and evasion maps). Approximately 75 per cent of the large print volume was of three particular combinations, namely 43A/B, 43C/D and 43KEast/West. The sheets specifically provide cover of Western Europe, including Denmark, Holland, Belgium, France, Spain, Portugal, Germany, Switzerland and northern Italy, and appear to have been printed in 1943 and 1944, prior to the D-Day landings. This comprehensive spatial coverage in itself adds strength to the theory that they were produced in such large numbers for operational reasons as well as for escape and evasion purposes.

The sheets are multi-coloured and are layered, demonstrating a high level of technical competence in printing so many colours on fabric and maintaining the colour register. They are all printed on man-made fibre (MMF): none appears to have been printed on silk. All are irregular in size and coverage, and appear to have been produced by panelling together sections of the International Map of the World (IMW) sheets, which would again imply access to the reproduction material held by D.Survey. Copies of all sheets still survive.

The IMW series was originally proposed in 1891 as a new world series by the German geographer, Albrecht Penck, drawing together all the mapping of colonial exploration made by individual nation states during the nineteenth century into a collaborative international series to face the challenges of the new century and for the good of all humanity. The series was significantly extended in its coverage during World War I, largely through the combined efforts of the Royal Geographical Society (RGS) and the War Office. By the end of that war, some ninety IMW sheets had been produced by the cartographers of the RGS, providing coverage of the whole of Europe, the Middle East and North Africa. It was this series which provided the source of this escape and evasion series and which was also apparently deemed reliable enough for operational use during World War II.

[SERIES 44]

In terms of its specification [Series 44] was identical to [Series 43]. Similarly, this escape and evasion series had no title or individual sheet names. All sheets had the prefix 44, believed to refer to 1944, their year of production. The escape and evasion series has been identified as [Series 44]. A detailed list of the sheets, which are all small-scale (1:1,000,000) and of the Far Eastern theatre of war, appears in Appendix 6. It is the only series of escape and evasion maps to provide coverage outside the European theatre of war and adjacent areas in Africa and the Middle East. Indeed, it is likely that the series was produced as an operational rather than an escape and evasion series. There is certainly evidence in the correspondence files of the time which seems to point to this being the intention. In a letter to the Director of Military Survey, Lieutenant Colonel W. D. C. Wiggins dated November 1944, the Survey Division of Headquarters Strategic Allied Command (SAC) South East Asia (SEA) set out preliminary

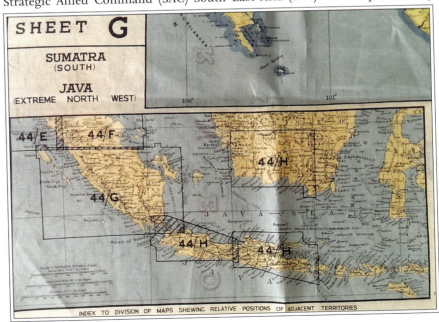

Adjacent sheets diagram for sheet 44G.

Part of Java around Batavia (now Jakarta) taken from sheet 44H.

enquiries prior to making a formal request. They wanted reassurance that if the SAC requested silk maps for use by assault troops operating in jungle and mangrove conditions, they could be provided in sufficient numbers. Wiggins replied on 15 November 1944 indicating that they would be better served by using US wet-strength paper or rag lithographic (RL) paper sprayed with cellulose varnish since these would be 'durable, waterproof and oilproof'.

Summary of [Series 44]

- ◊ 18 sheets identified
- ◊ 9 set combinations produced
- ◊ Coverage: South East Asia from Burma eastwards to China, Korea, Japan
- ◊ Scale: 1:1,000,000
- ◊ Print dates identified: 13 March 1944 to 4 October 1945
- ◊ Printed on: man-made fibre (MMF)
- ◊ Multi-colour printing
- ◊ Copies printed: over 185,000

For full details of the maps, see Appendix 6.

The series comprised eighteen basic sheets, produced in nine set combinations providing coverage of most of Indonesia, the Malay peninsula,

Thailand, Burma, French Indo-China, and the south and east central provinces of China. Again, the sheets were all printed on man-made fibre (MMF): none appeared to have been printed on silk. The sheets were all irregular in size and coverage, and are known to have been produced by panelling together sections of the IMW sheets. Diagrams showing the panelling of the IMW sheets to construct sheets 44N and 44O in this series appear in the files. Each sheet comprised sections of nine IMW sheets, which must have caused considerable production challenges. Copies of all sheets still survive.

[SERIES FGS]

This was the smallest group of escape and evasion maps. Five sheets have been identified in this series, produced as nine different combinations or single-sided. They are listed in Appendix 7 and are all small-scale, either 1:1,000,000 or 1:1,250,000 scale, covering the area of northern Europe and Scandinavia from Denmark and northern Germany to the northern extent of the Scandinavian archipelago, and eastward to the Finnish/Russian border. The sheets were of irregular shape and disposition, and again appear to have been produced by panelling from the IMW sheets since they were identical in specification to [Series 43] and [Series 44] above. The significance of FGS in the sheet numbers has not been identified, unless it was Finland, Germany, Scandinavia. Copies of all sheets still survive.

Summary of [Series FGS]

- ◊ 5 sheets identified
- ◊ Double-sided combinations produced: 4 identified
- ◊ Coverage: Scandinavia, adjacent USSR, north coast of Germany
- ◊ Scale: 1:1,000,000 and 1:1,250,000
- ◊ Print dates identified: 16 November 1942 to 21 April 1944
- ◊ Printed on: paper, fabric, rayon
- ◊ Multi-colour printing
- ◊ Copies printed: 105,150

For full details of the maps, see Appendix 7.

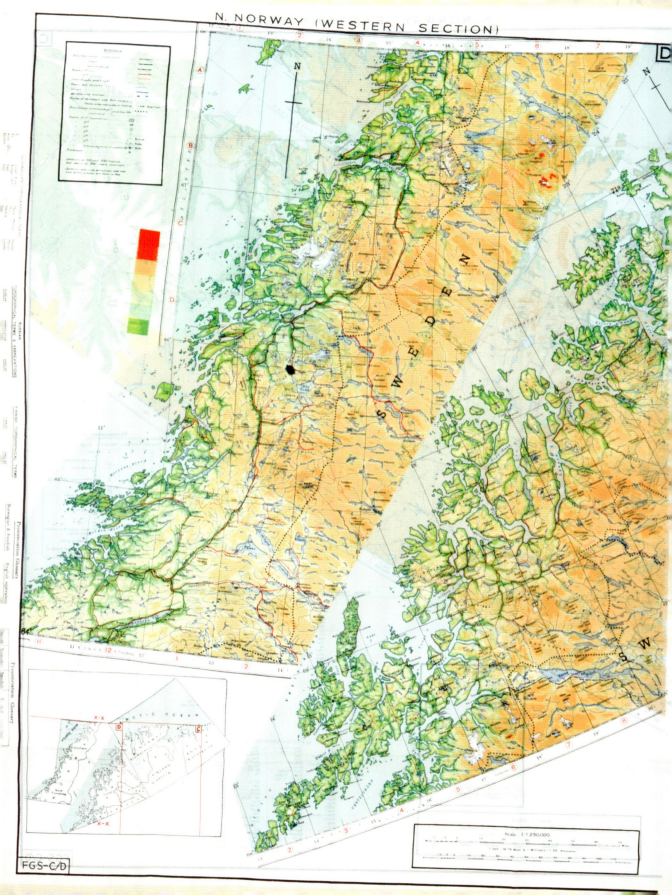

LEFT:
The irregular shapes and disposition found across [Series FGS] is clearly illustrated here with sheet D.

MISCELLANEOUS MAPS

The penultimate group was a collection of miscellaneous maps produced by MI9 apparently as briefing or reference maps. They were all small-scale and often based on existing maps produced for briefing and reference purposes by the Assistant Directorate of Intelligence (Logistics) since some of them carried the ADI(L) map reference. They are listed in Appendix 8. Some of the maps carry similar sheet numbers to those described in the Bartholomew series above, for

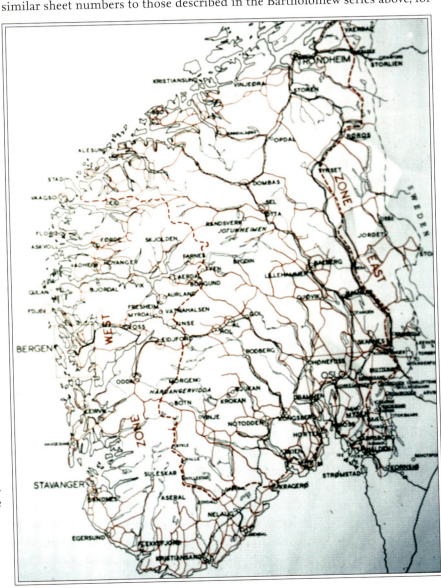

RIGHT:
Miscellaneous sheet Norway. The legend indicates that the map shows (in red) east and west zones garrisoned by German troops and where travelling without a permit is forbidden.

THE MAP PRODUCTION PROGRAMME 75

example 9J3 in this series and J3 in the Bartholomew series, yet another example of the confusion in MI9's cartographic procedures.

Summary of miscellaneous maps

- ◊ 16 sheets identified
- ◊ Double-sided combinations produced: 7 identified
- ◊ Coverage: Europe
- ◊ Scale: various
- ◊ Print dates identified: 14 January 1942 to 15 December 1943
- ◊ Printed on: tissue, Mulberry Leaf Substitute (MLS), silk, fabric, bank paper
- ◊ Copies printed: 20,050

For full details of the maps, see Appendix 8.

MI9 *BULLETIN* MAPS

A small group of maps was produced for inclusion in the MI9 *Bulletin*: these are shown in Appendix 9. In some cases evidence has been found that the maps were also produced on fabric for escape and evasion purposes. In other cases no copies on fabric have been identified. In at least one case, a map carries the same sheet number as an item in the Bartholomew series, namely A2, but is clearly a quite different map. This appears to be a further example of inconsistency in MI9's identification of their maps.

Summary of maps produced for the Bulletin

- ◊ 8 sheets identified
- ◊ Coverage: Schaffhausen Salient, Baltic ports, Norway
- ◊ Scale: 1:20,000 to 1:2,000,000
- ◊ Printed on: paper

For full details of the maps, see Appendix 9.

A surviving copy of the MI9 Bulletin. Its overall significance is discussed on page 86 and various maps from it are reproduced in Chapter 7.

THE PRODUCTION OF THE MAPS

The MI9 War Diaries indicate that production was initially managed on MI9's behalf by the Ministry of Supply working closely with commercial companies. The company which played a significant part in the production was John Waddington Ltd, Wakefield Road, Leeds, Yorkshire. Certainly, the remaining post-war stock of the maps, when located in a D.Survey map depot over thirty years later, was found to be interleaved with printed card from Waddington games. While the published history of the company made only passing mention of the important, covert, war-time role they played, the correspondence files of the time illustrate the almost daily contact which took place initially between Victor H. Watson and subsequently continued with his son, Norman V. Watson, the Managing Director, and E. D. Alston of Section CT6(c) in the Ministry of Supply located in Room 307 at 4 South Parade, Leeds. Alston was the acknowledged intermediary with the War Office section responsible for the requirement, i.e. MI9. There was certainly some direct, but very limited, contact between Waddington and MI9, the large proportion of the contact being with Alston as the intermediary. Although recognizing the security aspect of this separation and the need to keep the programme and its purpose a tightly controlled secret, there is no doubt that the lack of direct

Victor Watson photographed in his office in 1938.

contact, compounded by the lack of cartographic awareness by all the parties involved, created its own additional, and arguably unnecessary, challenges.

Waddington was apparently chosen because the company had already proved its ability to print on silk. The company was founded in Leeds in 1905. It initially specialized in theatrical printing and established a printing outlier in London. Later it diversified into general printing, playing card manufacture, board games and packaging products. Their expertise led to them being invited to produce the programmes for the Royal Gala (later Command) performance on silk for members of the Royal Family during the 1930s.

Christopher Bowes, a former Finance Director of the company, when depositing company archives with the British Library Map Library in 1999, posed the question about the company's wartime involvement with MI9: 'Why Waddingtons?' He offered the commentary that initially the company had not been 'very good' at printing the maps on silk and they had had to develop new techniques, adding that it might have been preferable for the work to have gone to commercial map publishers such as Bartholomew, Philips or even the Ordnance Survey who he suspected would have been better placed to do the experimenting. The point was well made. There is considerable difference between the printing in which Waddington was involved, i.e. board games, playing cards, posters, programmes, etc., and that of printing maps, even if they were afforded access to the original reproduction material for the maps. The challenge would have been exacerbated by the need to print the maps on silk. Bowes conjectured that there was some 'hidden politics' involved. He may have been right, but the character and approach of Hutton should also be considered. He was a man left very much to his own devices and not used to taking advice or guidance from anyone. There is no doubt that Hutton was convinced, wrongly, that his idea of maps on silk was original and unique. He was certainly single-minded in his approach, cartographically ignorant and sure that, because he knew and understood the objective of the exercise, he was correct in the approach he adopted.

The first recorded orders for the printing of the maps by Waddington are dated 6 January 1942. The following month the company purchased four sewing machines in order to machine-sew the edges of the silk maps. However, Hutton produced a booklet on 14 February 1942 in which he stated that, by that date, 209,000 maps had been distributed to units of all three Services, comprising fifty-six different maps on single-sided and double-sided silk or paper. Bowes regarded this as a rather odd statistic which could only be explained if, prior to that time, either other companies were involved in the printing or orders

Waddington's Wakefield Road factory in Leeds, which was demolished in 1992.

had been placed verbally with Waddington. There is, however, ample evidence that MI9, through Alston in the Ministry of Supply, was in regular contact with the company certainly from May 1941, and very possibly prior to that, and that considerable experimentation was taking place. On 12 May 1941 Alston wrote to the company commenting that the inks being used were apparently too old and 'did not work'. On 29 May he wrote again asking for the silk 'to be mounted' for printing. Any fabric on which an image was to be printed needed to be held taut during the printing process to prevent movement and the possibility of a blurred image. Certainly the usual method of printing on silk, and the one used by the silk printers of Macclesfield, which was a long-standing centre of silk-manufacturing in Britain, was to stretch the silk across a wooden frame.

By September 1941 there is evidence that Waddington was involved in secreting maps inside board games during the manufacturing process (see Chapter 4) and there is no doubt that the company was involved in printing the maps as well as hiding them inside board games in preparation for despatch to the prisoner of war camps. There are many examples in the company's files of Watson quoting map production costs to Alston and indicating that, for a range of reasons, the costs involved were up to 75 per cent above their usual production costs. On 6 January 1942, for example, he quoted for the cost of the:

> completed reproduction of Double Eagle, progressing with Emerald, costs for necessary photographic work, in preparing original, cutting material to size, preparing for lithographing, hemming on all four edges, backing sheets for mounting the fabric.

The Kodak company was involved in the production of the printing plates, although no detail survives. There are numerous items of correspondence detailing the number of copies required and the titles or sheet numbers of the items but, at no stage are they described as maps: they are rather referred to as posters or pictures. However, in all cases the sheet numbers and titles can be identified as items in the MI9 escape and evasion map production programme. A comprehensive three-page list of sheets in Series GSGS 3982 was produced around mid-February 1942. A list of nine maps produced on 26 February 1942 can be identified as Bartholomew-based maps, including the Caspian Sea (sheets T2/T4) and Kenya (sheet Q). Apparently the maps were often despatched by rail to London Kings Cross Station addressed to Major C. C. Hutton and marked 'to be called for'.

Henry Town of Allied Paper Merchants, located at 81 Albion Street in Leeds, was involved in experiments trying to increase the strength of the paper being used without increasing its thickness. At the same time the Ministry of Supply asked Waddington to carry out trials on Bemberg yarn, a man-made replacement fabric for silk and often referred to as Bemberg silk. It is also clear that academics were involved in various trials, Professor Briscoe of Imperial College, London, being mentioned by Alston in the Ministry of Supply in a letter he wrote to Watson on 7 April 1943, commenting on the 'Collodio–Albumen process' (used in preparing printing plates). It is also likely that academics at the nearby University of Leeds, which had a highly renowned textiles faculty, were consulted. Experiments were carried out, authorized by the Ministry of Supply, in printing methods and also into the use of different printing inks, the latter involving chemists from ICI visiting the Waddington factory in Leeds in June 1942.

SOURCING OF SILK AND PAPER

The sourcing of silk was a major issue for the Ministry of Supply, their priority always being for parachute manufacture rather than map production, and it is clear that there was a continuing search for a suitable fabric to replace silk. Supplies from the Far East (especially Japan), not unexpectedly, dried up and they had to turn to their contacts in the English silk industry based in Macclesfield for help in securing alternative sources. The leading Macclesfield silk manufacturing companies had had the presence of mind to stockpile raw silk from Japan prior to the outbreak of hostilities in the Far East at the end of 1941 but this supply was soon under considerable pressure from the Ministry of Supply. Nylon had only just been developed (in the USA) and was not yet available in the UK, at least not in the quantities required. All the standard supply sources for silk, China, Japan, Italy and France, were in enemy hands. It was necessary to identify alternative sources as a matter of urgency. That task fell to Peter William Gaddum. Born in 1902, he worked for the family silk firm of H. T. Gaddum from 1923, and, aged 36 at the outbreak of war, he was already serving in the Army. He was allowed to leave the Army and became Chief Assistant in the Ministry of Supply, responsible for the supply and control of silk and rayon. He proceeded to source silk from the Near and Middle East and made his first purchase in Cairo in December 1941. From there he travelled to Beirut in Lebanon, where he remained based until January 1944. Silk cocoons were available in Turkey, Iran and Lebanon, and much of the reeling of the silk took place in Lebanon. In the two years he remained in the Middle East, Gaddum travelled widely to ensure a secure supply of the silk needed at home for the war effort, always ensuring that he obtained the highest possible grade of silk. During the period, he visited Baghdad, Tehran, Cairo, Karachi, Delhi, Calcutta and Mysore. Eric Whiston, the son of another famous Macclesfield silk manufacturing family, was sent out to Lebanon to assist Gaddum, who eventually returned home in December 1944. Whiston moved to Rome in December 1944 and was based with the Allied Command.

While there is no evidence that any of the escape and evasion maps were actually printed in Macclesfield, Macclesfield silk manufacturers were involved in the finishing of the maps, i.e. hemming and folding. In addition to the continuing search for silk, it is clear that MI9 and its agent, the Ministry of Supply, was always looking for acceptable silk substitutes, since the priority for dwindling silk supplies was always parachute production. Once the USA had joined the Allies, the Ministry certainly made good use of recent US technical advances in the manufacture of rayon, a man-made substitute for silk first manufactured in the UK in 1905. Mulberry leaf pulp, which could be manufactured into an extremely thin and near-noiseless paper, variously referred to in the records as 'ML' (Mulberry Leaf) or more usually as 'tissue' or simply 'paper', was also used. The pulp had been discovered by the Royal Navy on a Japanese ship immediately prior to Japan entering the war. It proved to be a timely and valuable cargo which the ship's Captain was persuaded to give up.

RELATIONSHIPS WITH THE MILITARY MAP-MAKERS

It is clear that, certainly in the early part of the War, MI9 had little contact with MI4 and the military map-makers of the time, specifically D.Survey which was the military mapping organization responsible for the production of all operational mapping for both the Army and the Royal Air Force. Indeed, responsibility for the escape and evasion map production only passed to D.Survey, at that time located in Bushy Park, Middlesex, on 10 August 1944.

The pre-war removal of MI4 from London and its relocation in Cheltenham had caused increasing difficulties and concerns, since they were separated from the Operations, Intelligence and Planning Staffs at the War Office in Whitehall and also from the Air Ministry Map Section located at Harrow. Recognition of the need for daily contact with the General Staff in Whitehall and the need for bigger and better accommodation than was available at Cheltenham prompted the decision to move back to London in mid- to late 1942, despite the possible disruption of operational mapping work at a critical stage in the war. MI4 had been re-designated Geographical Section General Staff (GSGS) and was moved to Eastcote in outer London and to the nearby factory site at Hanwell, where the building was known as 'Hygrade'. At the same time the cross-service importance of GSGS was recognized and it was upgraded from Branch to Directorate level with the creation of the Directorate of Military Survey under Brigadier Martin Hotine. It was doubtless this move back to London which allowed MI9 greater access to existing operational mapping and to the production of escape and evasion maps based on existing GSGS series such as GSGS 3982.

A drawing produced by MI9 showing a proposed escape pack (see pages 102–104 for more details about escape packs).

4

SMUGGLING MAPS AND OTHER ESCAPE AIDS INTO THE CAMPS

'All my life, magicians, illusionists, escapologists in particular, have fascinated me. I expect it goes back to the night I tried to outwit Houdini.'
(*Official Secret* by Christopher Clayton Hutton)

The next challenge faced by Hutton was to devise a fool-proof system to deliver to those imprisoned in the camps the various escape aids and gadgetry which MI9 had contrived to have produced, not the least of which were the maps themselves. He later confessed that he spent many sleepless nights wrestling with this challenge. MI9 was acutely aware that most of the many thousands of men taken prisoner at Dunkirk had little or nothing by way of escape aids and devices and yet each of them was a potential escaper. They reasoned that getting the aids through in small numbers was not what was required; they needed to devise methods which would allow a steady flow of thousands of aids and devices. This was precisely what they attempted to do and it was here that the staff of MI9 were able, once again, to demonstrate their considerable ingenuity, resourcefulness and creativity.

THE ARTS OF SMUGGLING AND MAGIC

It is fascinating to realize the extent to which Hutton and his colleagues resorted to the basic tenets of smuggling which, over the centuries, British smugglers had developed into an art form. With a coastline of some 15,000 kilometres (9,300 miles), the United Kingdom's borders are impossible to patrol in their entirety and, over the centuries, determined smugglers have exhibited ingenious expertise in outsmarting the customs authorities. There is a considerable history of contraband being smuggled into the country through hollowed out containers, seemingly of standard construction when viewed externally, or concealed beneath the false bottoms of apparently standard containers. The significant characteristic of items designed to carry contraband goods was hollowness and it was a lesson which Hutton quickly incorporated into his plans. What MI9 embarked on in 1940 was thus essentially a programme of large-scale smuggling to ensure that the maps and other escape aids which they had produced would successfully reach the prisoners of war and thereby ensure that their planned escapes had the greatest possible chance of success.

Hutton confessed to his own childhood fascination with magic and the illusionary escapology tricks as practised by Harry Houdini, and the extent to which he had personally been involved in challenging Houdini (see page 21). In his personal reflections, Hutton provided a remarkable insight into his penchant for devising means to deceive the enemy. Few men could have proved better suited to the task. It was against this backdrop of his own fascination with smuggling and magic that the programme to smuggle escape aids into the camps began to take shape.

ROLE OF THE MI9 *BULLETIN*

Part of the training which Intelligence Officers received at MI9's Training School in Highgate was a briefing on the types of escape aids which had been devised. A key tool in this was the *Bulletin*, an extremely detailed volume compiled by MI9, which was essentially a textbook designed for use by Intelligence Officers when instructing others, and which was likely regarded as the escapers' handbook. It was produced as a small-format, loose-leaf manual which was easily updated through the regular issue of amendments, in numerical sequence, each carrying the month and year of issue. The overall security classification of the *Bulletin* was MOST SECRET. The first entry was a short, true story about Winston Churchill's experience of escaping in the Boer War when he dropped over the wall of where he was imprisoned by the Boers but his partner, who had the escape aids and maps, failed to get out: the moral of the story was emphasized: 'Always Carry Your Escape Aids With You.'

The *Bulletin* comprised twenty-three chapters in all, with the first seven covering such topics as security, escape and evasion, escape aids, international law, interrogation by the enemy, field-craft, and travel through Europe by rail. The subsequent chapters, each dedicated to a particular country or geographical region, provided a detailed description of the circumstances in each, recommended routes for escapes and detailed supporting maps. There were tips on successful escape ruses, modifying physical appearance (how to blend in) and recommended behaviour.

It was in Chapter 2 of the *Bulletin* that all the various aids to escape were listed and described in detail, but it was in the individual country chapters where copies of the maps produced to support particular, recommended escape routes, were to be found. Maps Nos. 4 and 5, Schaffhausen Salient (East) and (West), numbered A1 and A2 respectively, were in Chapter 15/Germany of the *Bulletin*. The significance of these maps is described in detail in Chapter 6 and details given in Appendix 9. The maps were accompanied by ground photographs of the border area showing detailed views of topographic features (streams and paths) and landmark features (telegraph posts). It was also in Chapter 15 of the *Bulletin* that a set of large-scale plans of the Baltic ports was to be found. These were detailed plans of Danzig/Plan No.1 marked A3, Gdynia/Plan No. 2 marked A10, Stettin/Plan No.3 marked A11 and Lübeck/Plan No.4 with no additional sheet number marked. None of the Baltic port plans was marked with a security classification. This absence of any security marking was notably unlike the two small-scale maps of Norway Military Zones marked Map B covering southern Norway, and Map C covering northern Norway, both dated March 1943, which were both classified SECRET. Details of all these maps can be found at Appendix 9 and the significance of the Baltic port plans is considered in Chapter 7.

Operational aircrews were only allowed direct access to the *Bulletin* under the supervision of their Commanding Officer or the RAF Station's Intelligence Officer. They were not, however, under any circumstances allowed to make notes or copy items in the *Bulletin*: everything had to be memorized. In the case of the maps, this must have been particularly challenging, not least since those being briefed had absolutely no idea of where they might find themselves in the event of capture. Notwithstanding the restrictions placed on its access and use, there is no doubt that the *Bulletin* contained the very best information and advice on escape and evasion known to MI9. By autumn 1942, it was being updated by the issue of regular monthly amendments which also included intelligence information gleaned from the reports of the debriefing interviews to which all successful escapers were subjected.

The *Bulletin* was indeed a veritable bible of escape and evasion, and contained everything that could be of assistance to Service personnel who found themselves cut off in enemy occupied territory or captured by the enemy. It was devised as a textbook for all who were called on to give instruction on escape and evasion and its contents could change regularly and radically in the light of new information received from the field.

THE USE OF BOARD GAMES, SPORTS EQUIPMENT AND OTHER LEISURE ITEMS

Hutton and his team designed a great variety of devices to get material to prisoners of war. Maps and other escape aids were smuggled into the camps hidden inside all sorts of items such as games boards (Monopoly, Ludo, Snakes

An MI9 design for a game, with a hidden compartment for a map, currency, saw and compass.

and Ladders, draughts), dart boards, cribbage, backgammon and chess sets, pencils, gramophone records and sports equipment such as table tennis sets and squash rackets.

It has been suggested that cricket bats and balls were also sent into the camps, although no proof has so far been discovered that these also conveyed escape aids. Certainly there exist some family history stories that companies such as A. G. Spalding & Bros. were manufacturing cricket bats with silk maps hidden in the handles and compasses in the top of the handles. John Worsley was a war artist who had been captured and sent to the Marlag and Milag (Marinelager and Marineinternaten Lager) Nord camp constructed by the Germans southwest of Sandbostel, some forty kilometres (twenty-four miles) northeast of Bremen. One of his surviving watercolours depicts a cricket match in full flow. While it is difficult to make out the bat easily, the stumps and wicket keeper, complete with pads, are very clear, providing evidence that cricket equipment was sent to the prisoners of war. Indeed, there are also mentions of cricket matches in some of the reminiscences of prisoners of war. There is, however, no indication in the records that A. G. Spalding was ever involved in covert activity with MI9.

A cricket match at Marlag and Milag Nord camp, painted in 1944 by John Worsley, an official war artist, who was imprisoned there. It has been suggested, but not proved, that cricket bats were used to smuggle escape aids into camps.

Board games

Hutton tackled this very challenging part of his task with apparent alacrity. The first company which he turned to was John Waddington Ltd of Wakefield Road, Leeds. He was already in regular contact with the company, through the Ministry of Supply, having established that they were rare, if not unique, in the commercial sector in being able to print on silk. Since they were already printing maps on silk for MI9, it was but one further step to ascertain the extent to which they could help MI9 in the manufacture of the means to despatch the maps into the camps. The fact that they were the principal agent in the UK to manufacture and distribute the British version of the popular US board game Monopoly was an undoubted gift to MI9.

A wartime copy of Monopoly. Some boards had escape maps hidden within them, while genuine local currency was sometimes smuggled within the pile of Monopoly money.

Monopoly boards began to be manufactured with escape maps hidden inside them. Those containing maps of Italy had a full stop after Marylebone Station and those containing maps of Norway, Sweden and Germany had a full stop after Mayfair, as a letter dated 26 March 1941 from Victor Watson of Waddington to Christopher Clayton Hutton explained. Additionally, boards were also produced containing maps of northern France, Germany and associated sections of the frontier: MI9 again insisted they carry some distinguishing mark and, in this case, it was a full stop after 'Free Parking'. The practice ensured that there was a ready means of differentiating which boards carried particular maps and this system of coding also ensured that the appropriate games were sent to prisoners of war in the appropriate geographical area. Local currency was, of course, also needed by those planning to escape and the Monopoly money provided an ideal hiding place for local currency. Waddington was also sent in March 1941 a selection of small metal instruments, such as saws and compasses and asked if they could consider how best to secrete these in the Monopoly sets.

Playing cards

Waddington was also a major manufacturer of playing cards. In 1999, Christopher Bowes, the Waddington company archivist, described how maps were smuggled to prisoners in a pack of cards:

> The map would be made from a material impervious to water and sandwiched between the back and the pip side of the cards by water soluble glue. When dropped into a bucket of water the cards would come into three parts: front, back and the map. Each segment of map was serially numbered (in orange on the cards that I have seen) and overlapped the next segment on all sides by about ¼ inch. Each pack contained one map in 48 segments. The four aces contained a small-scale map of Europe and the Joker held the key or map legend. To make up such a pack must have been fiendishly difficult because of the overlaps.

On 11 May 1942 a letter was passed to the company from the Ministry of Supply with an order for 224 packs of playing cards, twenty-four packs of which were to be Prize Packs 'on the formula we discussed': this appears to relate to the type of packs described by Bowes. It is clear from ensuing correspondence that a considerable increase in the cost of manufacturing such playing cards resulted, not least because of the need to ensure adequate overlaps between the cards. The overlaps were deemed necessary to ensure the total map was covered and that no portion would be missing. Such increased cost was later queried

A pack of cards with map sections sandwiched between the front and back of each card and a games box containing a secret compartment.

as being 'rather high' and the statement was made that the requirement was deemed 'not of sufficient importance to warrant expenditure'. This was a rare, possibly unique, indication that MI9 was alert to the costs of its endeavours. The cards which were produced were packed into individual packs and also in bridge sets.

The *Per Ardua Libertas* volume was produced by Hutton in February 1942. It was a detailed photographic review of the range of work for which MI9 had been responsible during the first two years of its existence. The volume was bound in red leather and was clearly a high quality presentation piece, possibly seeking to influence their sister organization in the USA, MIS-X, the Americans having declared war on the Axis powers on 11 December 1941. The volume showed a photograph of how maps were hidden in playing cards. It also contained photographs of various sizes of chess sets and confirms that maps were hidden inside them during manufacture. Norman Watson of Waddington was able to provide Alston, his principal contact in the Ministry of Supply and intermediary with MI9, with details of three companies in London which marketed pocket chess sets, indicating that they were probably agents rather than actual manufacturers. The companies were Thomas Salter, Baileys and E. Lehman & Co. There is also separate confirmatory evidence, in the form of the original artwork held in the RAF Museum, that chess sets were produced with secret compartments inside the boards, within which maps could be hidden (see page 124).

A page from Per Ardua Libertas *showing some of the smuggling methods used by MI9. The example at the top of the page shows a pack of playing cards containing a map. Also shown are cigars which here are described as being used to smuggle maps and compasses but which were also used to send a message of support from Winston Churchill to prisoners of war (see page 196).*

Gramophone records

During the course of the war, Hutton needed continuously to generate new and ingenious ideas for hiding the maps and other aids, not least because the German camp guards did discover some of his early hiding places. Hutton acknowledged that the majority of the hiding places were undiscovered for long periods of time but each, in turn, was eventually spotted by an alert guard and had to be replaced by a new host device. Once MI9 had been alerted to the discovery, they suspended distribution of the items until an alternative replaced it.

Initially, for example, Hutton had used books but, once they had been discovered as the repository of escape aids, he turned his attention to gramophone records, approaching John Wooler, Head of the Record Development Laboratory at EMI. Hutton's initial idea was apparently to conceal miniature compasses inside the records. However, he rapidly discovered from Wooler that it was also possible to conceal maps and currency in the records. Wooler pointed out to Hutton that when the Columbia and His Master's Voice (HMV) record companies amalgamated in 1930, the two companies were using very different processes for record production. Essentially, HMV records were solid, whereas those produced by Columbia used a more economical method of Col Powder lamination (powder adhesives were used in the lamination process

A gramophone record of Beethoven music conducted by Arturo Toscanini that contains secreted maps. Later on in the war MI9 also hid maps inside the record cover.

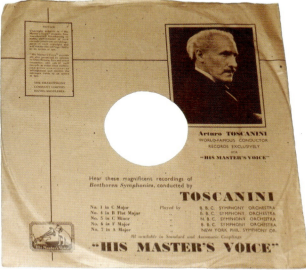

as an alternative to heat lamination). Hutton quickly realized the possibility of exploiting the lamination process since it allowed items to be hidden under the laminate in specially incorporated compartments. By adding extra layers it was possible to conceal up to four maps in each record or a combination of maps and currency. It is clear that Wooler did the pressing of these special records himself at weekends when the usual press workers were not around. Records were packed into boxes, each box containing five records.

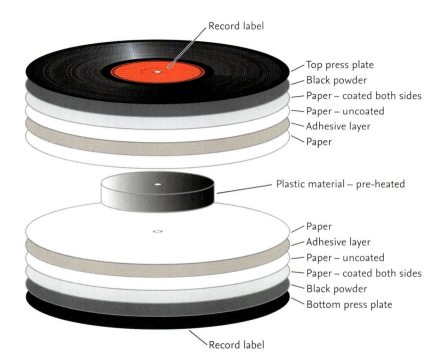

Drawing, based on an indistinct manuscript sketch in the RAF Museum, which shows the layered construction of a gramophone record. Notice the labelled paper layers towards the bottom of the diagram, showing where maps and currency could be secreted. The same layers are repeated near the top of the diagram, enabling up to four items to be included in each record.

One particular record contained a frontier map in two parts, i.e. the map and the route. While not specified, the map is believed to be the Schaffhausen map, sheet Y (see Chapter 6 and Appendix 1). Evidence survives of the despatch which took place on 14 May 1941, and it is clear that some of these boxes were destined for Oflag IVC (Colditz), Stalag XXI D (Posen) and Stalag Luft III (Sagan). Sagan can only be identified by four men whose names appear on the despatch list, namely Flight Lieutenants J. C. Breese and D. A. Ffrench-Mullen, Pilot Officer W. H. C. Hunkin and Squadron Leader W. H. N. Turner, and whose names also appear in the Sagan camp history as coded letter writers. Certainly it is the case

TOP:
A letter from Hutton to HMV with some packing instructions for the records (for an explanation of Red and Black, see illustration below). This letter has a rare example of Hutton's signature.

BOTTOM:
Part of a despatch note for gramophone records sent under the cover of the Prisoners' Leisure Hours Fund (abbreviated to P. L. H. F. at top right). Capt. P. R. Reid at Oflag IVC (Colditz) is listed as one of the recipients; he was later to escape from Colditz (see Chapter 6).

that some of the individuals to which the records were despatched subsequently escaped successfully, for example Captain P. R. Reid from Oflag IVC (Colditz). The timing of the despatch is also such that the hidden map might have helped in the escape from Colditz of Airey Neave (see Chapter 6).

96   GREAT ESCAPES

A total of 1,300 gramophone records containing maps were produced and despatched. Hutton's choice of the particular gramophone records to send was clearly subjected to the same careful consideration and reasoning which was the hallmark of all his work: he chose to send recordings of the works of Beethoven and Wagner, studiously avoiding the selection of any Jewish composers which would have been automatically confiscated by the German censors or guards. While the records were perfect in every aspect of manufacture and could be played, Hutton did see the irony of the fact that the prisoners of war had to break the records in order to access the concealed items and, employing a touch of black humour, he dubbed the whole enterprise 'Operation Smash-Hit'.

Christmas crackers

It is clear that other members of MI9 came up with ideas for concealment of escape aids from time to time. One ruse was apparently the brainchild of Jimmy Langley (see page 23). His idea was to hide maps, currency, compasses and dockyard passes inside Christmas crackers which were despatched in September 1943 in boxes as apparently innocuous seasonal treats from the Lancashire Penny Fund, one of MI9's cover organizations, to arrive in the camps in time for Christmas. The parcels were accompanied by an open letter to the Camp Commandant inviting them to share in this harmless Christmas cheer. The prisoners of war had been alerted in coded letters beforehand which colour box

A Christmas celebration at Oflag IVC (Colditz), as recorded by an official German photographer. Crackers would not have been out of place with the Christmas tree and streamers.

was 'good' and which was 'naughty' (MI9's terms) to ensure that only the innocent crackers went to the German captors, although it was subsequently reported that the escape aids in some of the crackers were discovered by the Germans. The matter was apparently regarded as serious enough for the Germans to report it as a breach of the Geneva Convention. Under the Convention, the Red Cross held a unique position and was allowed access to prisoners of war because of its impartiality and neutrality. However, since the 'contraband' had not been sent under cover of the Red Cross, the referral apparently resulted in no action.

Other festivities were exploited as well. In October 1943, as part of a Hogmanay scheme, sticks of shaving soap were sent in toilet parcels into eight camps where there had previously been no contact with MI9. The sticks of soap contained maps, compasses, money and a message. It was successful in four of the eight camps chosen.

Pencils

Pencils were also used as a secret conduit for the provision of maps. Certainly at one stage the Cumberland Pencil Company became involved in covert activity with MI9. The company is mentioned in a post-war list of companies as being involved with MI9. The Pencil Museum in Keswick has examples of pencils that appear perfectly normal and usable but which contain a silk or tissue map rolled very tightly inside the pencil in place of part of the pencil lead. In addition a miniature compass was hidden under the rubber at the top of the pencil. The

Cumberland pencils, now in the Pencil Museum, Keswick, showing the map and compass that would have been secreted within them.

pencils were distinguished by being painted dark green (as an economy, wartime pencils were unpainted) and the number stamped on the pencil (103 in the example illustrated here) indicated which map the pencil contained, although no records remain that link the codes to specific maps. *Per Ardua Libertas* also contains an illustration of maps being concealed in bridge marker pencils.

OTHER ESCAPE AIDS: COMPASSES AND PURSES

From the beginning Hutton had decided that compasses were as vital as maps to aid the escapers. He designed them in a quite bewildering variety of forms. Miniaturized, they were hidden in the buttons of RAF uniforms, on both trousers and tunics. When eventually the buttons were found by German guards to hide compasses, Hutton simply resorted to altering the screw direction by having them manufactured with a left, rather than right, hand thread which proved to be a simple but very effective change. In addition, Hutton arranged for almost anything made from metal to be magnetized, for example razors, hacksaws and pencil clips. Together with maps, compasses were sometimes hidden in a small compartment inside the heel of an RAF flying boot so that any air crew who were shot down, and managed to evade capture, had at least a fighting chance of finding a route to escape successfully.

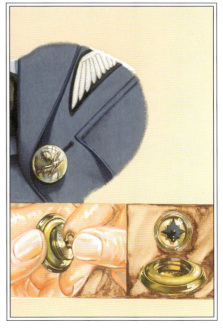

LEFT:
A button compass and a selection of other escape aids, including a miniature telescope.

RIGHT:
An MI9 drawing of a compass hidden within a button on an RAF uniform.

Purses with currency and maps were also provided for RAF flight crews overflying occupied Europe. They were essentially tobacco pouches but they were also used to hold maps. The term Purse Maps was at one point coined as the generic term for all escape and evasion maps produced on fabric and tissue. Some years ago, in the early 1980s, the Intelligence Corps Museum held a volume of escape and evasion maps entitled Purse Maps. It was a large, black-bound folio of one hundred pages with 'PURSE MAPS' marked on the spine. It contained maps on thirty-seven pages, many of them double-sided and a few of which have never been found elsewhere in other collections. The source of the folio in the collection at that time was clearly indicated as being MI9.

Purses were also produced containing town plans and addresses of British Consulates in Spain, although there is evidence that these were in fact photographic plans which were subsequently considered to be undesirable for inclusion. MI9 directed that they should be removed and destroyed as a new type of purse was to be made available and the older ones were to be withdrawn and returned to MI9.

LEFT:
A fabric tobacco pouch containing an escape and evasion map with France on one side and Germany on the reverse.

RIGHT:
This Purse Map showing France (sheet 9Ca, see Appendix 8) was issued to Flight Lieutenant John H. Shelmerdine, DFC who, for three years from April 1942, flew Spitfires out of RAF Benson in Oxfordshire on photographic reconnaissance flights over Western Europe. The red shaded area identifies the 'Coastal Defence Area', which was much more militarized than the rest of France. Note the number of folds that were required to fit the map into a purse.

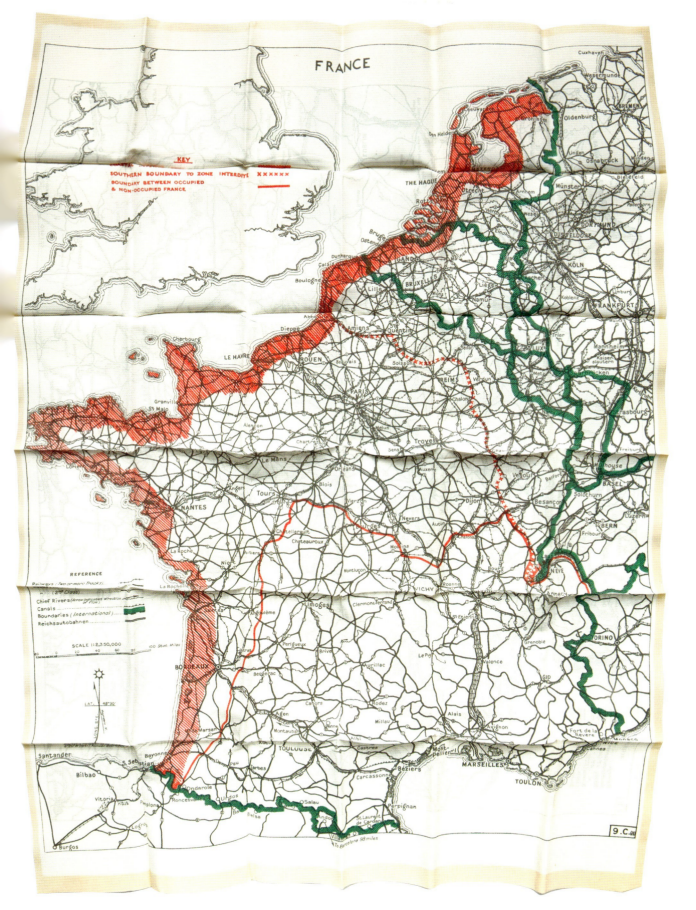

ESCAPE PACKS

Johnny Evans (see page 21) had long advocated to Hutton the need for some form of food-pack or emergency ration box to sustain the evader. Based on his own escape experience in World War I, Evans assured Hutton that 'the escaper's greatest enemy is hunger. When a man is starving, he very soon becomes reckless and insensitive.' In fairly typical fashion, Hutton initially decided against involving the Quartermaster's department in whose field of responsibility this would fall, but rather went directly to commercial companies who had proved to be so helpful and supportive thus far in MI9's covert work. He felt that the standard fifty cigarette tin was ideal in size and shape as it would fit into the breast pocket of an RAF flyer's uniform and also into the map pocket of battle-dress trousers. He acquired these cigarette tins in large numbers from the W. D & H. O. Wills cigarette factory in Bristol, having personally approached the company's directors to seek their cooperation.

An MI9 drawing of a stopper for a water bottle that could double up as an escape pack. The stopper contains a hidden compass and watch.

After experimenting and listening to the experiences of former escapers such as Evans, Hutton apparently decided that each pack should contain malted milk sweets (provided by Horlicks), chewing gum, a bar of peanut blended food, water purifying tablets, a rubber bottle for water, a small saw, a bar of chocolate, Benzedrine tablets (for the purpose of keeping escapers awake at critical times), matches, a compass, thread, tape and two tissue maps (one of Germany and one

Sketch showing the contents of a standard escape pack.

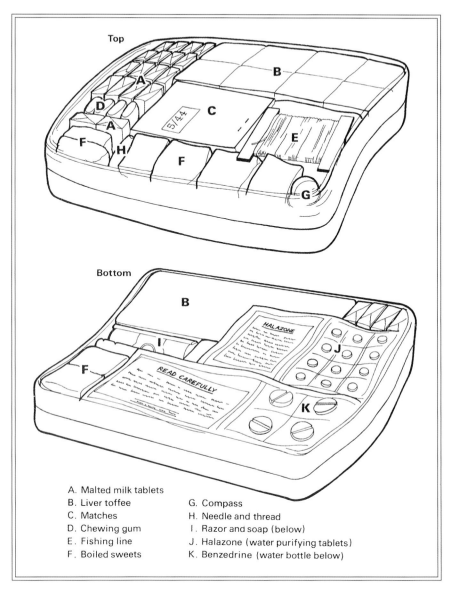

A. Malted milk tablets
B. Liver toffee
C. Matches
D. Chewing gum
E. Fishing line
F. Boiled sweets
G. Compass
H. Needle and thread
I. Razor and soap (below)
J. Halazone (water purifying tablets)
K. Benzedrine (water bottle below)

of northern France). This escape pack became standard issue for RAF crews overflying occupied Europe. Later, after the Chairman of Halex, the toothbrush manufacturers, had been persuaded to help, a new case was manufactured in plastic. This plastic case had three advantages over the cigarette tin: it was waterproof and transparent, it had a compass built into the screw top and it could be moulded to better fit the human frame, proving less obtrusive and rather more comfortable than its predecessor. Indeed, the escape pack became very popular. It was later modified further to avoid the inclusion of a rubber bottle, rubber having become scarce, whereby it was made in one piece of plastic with a screw top to serve as a water bottle in its own right. Despite preferring to avoid the 'pettyfogging bureaucrats' in the Ministry of Supply, Hutton found that he had to deal with them since he needed food supplies to ensure the efficacy of the escape packs. He had a number of brushes with them but largely succeeded in acquiring supplies of everything needed for inclusion in the escape packs.

As with some of the maps, after the war ended, the Ministry of Supply sold off some of the surplus escape packs. It is clear that some of these were purchased by an enterprising individual who subsequently advertised them for sale as 'Holiday Accessories for Campers' at the not inconsiderable price at the time of 11/- (shillings), the equivalent of around £13.60 today.

DELIVERY AND COMMUNICATION

Having created a diverse range of escape aids, there was a pressing need to develop a reliable system of delivery. They had to smuggle the items into the camps without arousing the suspicions of the enemy. There were two existing supply channels, both provided for under the Geneva Convention, namely Red Cross parcels and monthly parcels sent by family and friends. MI9's system had to be different and separate to ensure that neither of these two existing channels was ever compromised. Such compromise would have inevitably resulted in the withdrawal of the privileges and MI9 was already only too aware of the extent to which some of the camps were dependent on Red Cross food parcels as a veritable lifeline since camp rations were often extremely thin. In response, therefore, distinct but entirely fictitious cover organizations were set up, amongst which were the Prisoners' Leisure Hours Fund, 'a Voluntary Fund for the purpose of sending Comforts, Games, Books, etc. to British Prisoners of War', the Lancashire Penny Fund and the Licensed Victuallers' Sports Association. The War Diary entry for August 1941 noted:

> This month has marked an important step forward in the parcel side of our work. In March, April and May large numbers of parcels containing escape material as well as clothing of all kinds were sent to prisoners under the auspices of 'The Prisoners' Leisure Hours Fund'. Considerable anxiety was felt as to whether the Enemy would swallow this Fund and it has been most gratifying for us to receive during this month no fewer than 63 acknowledgements of parcels despatched under its name.

These organizations had properly headed notepaper and apparently real, but entirely fictitious, addresses in London. The notepaper for the Prisoners' Leisure Hours Fund was particularly notable; not only did it contain details of the company's officers and non-executive Board members, but it also carried a quotation from Runyan (sic) 'the treasures to be found in idle hours – only those who seek may find', clearly an allusion to what was contained in the parcels and an apparent play on the work of John Bunyan's, *The Pilgrim's Progress*. It provided yet another example of the black humour which sometimes characterized MI9's covert work. There were many more of these cover organizations carrying such

PRISONERS' LEISURE HOURS FUND

"The treasures to be found in idle hours—
only those who seek may find."
Runyan.

President:
B. ATTENBOROUGH, Esq.

Vice-Presidents:
Sir THOMAS BERNEY, Bart.
L. C. UNDERHILL, Esq.

Committee:
Lady D. BROWNE,
The Hon. Mrs. E. FREEMAN,
P. O. NORTON, Esq.
J. B. WORLES, Esq.

**66 BOLT COURT,
FLEET STREET,
LONDON, E.C.4.**

Hon. Treasurer:
E. TOWNSEND, Esq., C.A.

Hon. Secretary:
Miss FREDA MAPPIN.

Telephone:
CENTRAL 3951

12th MAY, 1941

Dear Sir,

    Through the kindness of one of our contributors, we are enabled to send to you a selection of Musical Instruments - and Gramophone Records, and we are having despatched direct from the manufacturers in the course of a few days some records.

    We intend despatching different selections for each prisoner of war - to whom we send these, and it is hoped in order that all may enjoy the variety, you will offer to interchange with each other.

    Further supplies will be sent you at regular intervals, and if there is any particular record you desire sent, perhaps you will look through the Catalogues we are sending letting us know the make and number, and we will do our best to despatch to you in due course.

    Trusting you are enjoying good health, and looking on the bright side of things.

Yours faithfully,

Secretary.

A Voluntary Fund formed for the purpose of sending Comforts, Games, Books, etc. to British Prisoners of War.

A letter from the Prisoners' Leisure Hours Fund, advising the recipient of the arrival of gramophone records, as reproduced in Per Ardua Libertas.

titles as the Authors' Society, Browns Sports Shop, Jig Saw Puzzle Club, League of Helpers, The Old Ladies Knitting Committee, The Empire Service League, Crown and Anchor Mission and others with similarly innocuous titles.

MI9 was keen to test the extent to which parcels were getting through successfully to the prisoners in the camps and, therefore, inserted special cards requesting that the Camp Commandants allow the prisoners of war to acknowledge receipt, thus saving the prisoners one of their weekly letters. The Germans proved to be very helpful in allowing the prisoners to send back their signed receipts. To afford as much cover as possible to the 'naughty' parcels, MI9 also sent in 'good' parcels containing much needed winter clothing, jumpers, socks, vests and the like. The receipts clearly indicated what had been received and were franked with the stamp of the oflag from which they had come. A number of the surviving examples show clearly the stamp of Oflag IVC (Colditz).

LEFT:
A receipt of a parcel from the Prisoners' Leisure Hours Fund, stamped with the Oflag IVC (Colditz) stamp.

RIGHT:
A receipt for five gramophone records signed by Flight Lieutenant J. C. Breese on 6 August 1941 and sent back from Stalag Luft III (Sagan). He was sent the 'naughty' parcel 12310 on 14 May 1941, according to MI9's records.

There is no doubt that the Germans did find some of the escape aids and the prisoners did try to alert MI9 to any mishaps. In the coded letter dated 14 February 1943 from John Pryor, the hidden message was:

GERMAN FOUND CONTENT OF GAMES PARCEL FOR FRANKLIN MILAG

Franklin would have been the name of the recipient and MI9 would know what the parcel contained and, therefore, which method of hiding the escape aids had been compromised. Such a warning would have ensured that that particular method would not have been used again and no further parcels would have been despatched to Franklin. It has already been noted in Chapter 1 that the Germans were aware of at least some of the escape and evasion maps, through discovering them in clandestine parcels, and the extent to which escape was regarded as a duty by the British officers and men.

The men in Stalag Luft III (Sagan), as noted in its camp history, regarded the most secure method of concealing escape aids to be:

> in games parcels, especially in poker chips, shove ha'penny boards and dartboards, chess sets etc. A first rate carpentry job of this kind could not be detected unless the article was destroyed completely.

Apparently every method of concealment was successful at one time or another: the German censors tended to concentrate on one category for a period, as their suspicions were aroused, and then turned their attention to another category, i.e. all gramophone records would be searched before they moved on to search all items in another category. After the winter of 1942, all gramophone records which arrived in Sagan were apparently sent to Berlin to be X-rayed and 20 per cent were found to contain escape aids. No records were allowed in the camp thereafter until the Senior British Officer eventually persuaded the Camp Commandant to allow the German Medical officer to X-ray them, presumably as he knew that they were 'clean' by that time.

Certainly in the Sagan camp, all parcels arrived at the local railway station and the post office. They were collected by prisoners of war supervised by German guards and parcel censorship staff. Red Cross parcels were stored in one store and all other parcels were placed in a separate store where they were sorted by the prisoners of war who had been briefed by their Escape Committee, as a result of coded messages from MI9, on which parcels to select and remove. Cigarette and tobacco parcels were usually not checked by the Germans, as together with Red Cross parcels, they were regarded as above suspicion. The selected parcels were often, therefore, secreted into the sacks of the cigarette and tobacco parcels in order to smuggle them into the compounds. Games parcels were often subsequently smuggled back into the store, once the aids had been extracted, so that suspicion would not be aroused.

MI9 was in direct communication with the camps from the very earliest months of its operation. Initially, the only means of communication with the prisoners of war was through coded letter traffic. By 1942, this coded communication had developed markedly with messages being distributed more widely. As a result, escape equipment was being sent in increasing volume into the camps. By Christmas 1942, MI9 was feeling bold enough to send the first bulk parcel containing only escape material into Oflag IVC (Colditz). They were advised by a successful escaper from Colditz, likely to have been Airey

Three British prisoners of war chat to a German officer in the Red Cross parcel store at Stalag Luft III (Sagan). Piles of Red Cross parcels can be seen stacked up behind them. MI9 scrupulously avoided using Red Cross parcels, acutely aware that absolutely nothing should impede the flow of such parcels into the camps or compromise the Red Cross.

Neave, who had returned to the UK earlier that year, that they should send advance notification to the camp so that the store room could be broken into and the parcel's contents extracted. The plan worked successfully and the plan was implemented with other camps. Some 70 per cent of these bulk parcels subsequently successfully reached the camps and the men for whom they were intended. This considerably eased the problem of continuously searching for new ideas to conceal the escape aids. By August 1943, large quantities of escape aids were arriving in the camps, including material as bulky as cameras, typewriters and wireless sets. The wireless sets allowed far more immediate contact with the camps.

The degree of contact with the camps varied enormously. Stalag VIIA (Moosburg), for example, did not apparently receive very much by way of escape aids. This camp was located some forty-seven kilometres (twenty-nine miles) northeast of Munich and by 1944 there were in excess of 10,500 British prisoners of war in the camp. After the war, one of the liberated prisoners reported that the only escape aids they had received were contained 'in some gramophone records' which arrived in July 1942 and no further parcels were received for the duration of the war. This clearly indicates that delivery was patchy to some of the camps.

The methods of communication, especially the coded letter traffic, between MI9 and the camps, played a very significant part in the whole story of the escape and evasion mapping programme and are, therefore, addressed in detail in the following chapter.

CAMP ORGANIZATION

Given the considerable level of organization, planning and activity which characterized the work of MI9 to support successful escapes, it arguably comes as no surprise to discover that this was matched by a similar level of organization and activity in the camps. Each camp sought to create its own Escape Committee, headed by the Senior British Officer (SBO), or the Senior American Officer (SAO) in camps where the two nationalities were mixed. All escape plans had to be authorized and overseen by this Committee. While the organization, of necessity, was rather more loose than that of any operational battalion or squadron, organized it most certainly was, albeit in a very unobtrusive way. In the Marlag and Milag Nord camp, the Escape Committee had a system of 'patents'; all schemes were registered with the Committee and each person was allowed the first attempt with their own idea. The Escape Committee directed and supervised all activity in support of any plan it authorized, including the provision of the requisite escape aids. The aids provided by MI9 were carefully allocated, plans were monitored, and all supporting activity, such as the provision of civilian clothes, the forgery of local papers required for travel or the copying of maps, was overseen by the Committee.

It is likely that MI9 was helped in its attempts to hoodwink the Germans by the age, physical condition and lifestyle of the guards. Many of them were elderly and certainly not fit enough for operational action on the front lines. Their own lifestyle was far from comfortable, with the result that many were easily bribed. In some camps it is clear that the black market was extensive and the Germans were very corrupt, 'there being nothing that could not be obtained at a price for either money or cigarettes,' as James remarked. While some of the ruses and methods employed by MI9 were discovered and they had to keep coming up with new ones, it is clear that the sheer volume of activity by MI9 in keeping up the flow of escape aids helped to support many successful escapes.

Those prisoners who had particular expertise from their civilian employment were given support work which maximized the value of their contribution. Prisoners in the Brunswick camp, for example, created a crude, but very effective, print works to reproduce maps (see Chapter 8). John Pryor, whose personal

OVERLEAF:
A contemporary War Office map showing the location of all the prisoner of war camps in Germany.

OVERLEAF (INSET):
Map detail of the area around Colditz.

story will be recounted in more detail in Chapter 7, was regarded as the master forger in Marlag and Milag Nord camp. His son recalls that his copperplate handwriting could look as if it had been typeset. Clearly an innate talent was put to good use during his captivity in producing forged passes for potential escapers, an artistic talent which he subsequently employed to good effect as a hydrographic surveyor when he returned to active duty after the war. Those who had been tailors were employed to make the civilian clothes; those with any kind of technical, or even criminal, bent were able to pick locks, make keys or even steal items which were needed. The men had sufficient leisure time to scrounge or steal from, or bribe their captors; any bribery was usually done with cigarettes.

It was usual, though by no means consistently done, for those planning to escape to be employed on map copying, as this afforded them the opportunity to learn the geography of the terrain over which they planned to travel. There is, however, little mention in the subsequent escape reports of the maps, which appears to reinforce Professor Foot's comments in a conversation in 2012 that all who were issued with escape maps were briefed never to mention them: it is clear that few did. There is evidence that in some of the camps, roster systems and timetables were organized so that, as some of the men were deployed copying maps or producing forged documents, for example, others were deployed to act as lookouts and give warning of the approach of any of the German guards.

It is clear that there was a level of organization in many of the camps which allowed the escape activities to proceed relatively unencumbered. The infrastructure that was created allowed many of the escape plans to proceed to a successful outcome. It was very important in the greater scheme of things that the camp organizations were effective in realizing the benefits of MI9's detailed work in support of escape activity.

VOLUME OF PARCELS DESPATCHED TO THE CAMPS

MI9 kept detailed records of the despatch of parcels to the camps. Almost certainly this would have been in the form of a card index noting the contents of the individual parcels, to which camps they were despatched, when and what acknowledgements were received. Although this record does not appear to have survived, it is possible to track the volume of the parcels despatched through the monthly entries in the War Diary. For example, the entry for August 1941 indicated that 497 had been despatched up to that point, of which 78 had been acknowledged. An additional 193 clothes parcels, i.e. straight parcels, had also been despatched, of which 33 had been acknowledged. By July 1943, almost 400 parcels were despatched in a single month to camps in Germany and Italy and one of the parcels sent to Germany consisted entirely of escape aids. No evidence has been found to support the hypothesis that the Germans had an equivalent organization and there appears to be no indication that there was any similar traffic passing in the opposite direction, although it is not possible to make an unequivocal statement to this effect.

The story which emerges from this chapter is one of breath-taking ingenuity and inventiveness engendered by necessity, initially by MI9, but the more so by the prisoners of war themselves. Having embarked on a programme to produce escape and evasion maps in an appropriate form and in sufficient numbers to ensure that the escape philosophy was enacted in reality, MI9 sought to ensure that those maps reached the camps in prompt and effective ways. Escape, whether ultimately successful or not, took a great deal of planning both at home and in the camps, and MI9 showed their acute awareness of this and the extent to which the maps were a key part of successful escape. It is also clear that the prisoners of war showed admirable courage and indefatigability, apparently accepting and actively responding to the philosophy of war that it was each man's duty to attempt to escape. The Escape Committees must have been reassured in their commitment and endeavours when they saw the extent to which MI9 was seeking to support them. When they asked for items, MI9 did their best to respond. The escapes from the camps were always a team effort and for every man that made it home successfully, there was a support team left behind in the camps. Similarly, at home, there was a very real sense of teamwork and cooperation as commercial companies in the UK contributed both materially and in terms of their expertise to the whole escape and evasion enterprise which MI9 had created and fostered. It is certainly the case that the reality of what was done has proved to be decidedly more astonishing and impressive than that conveyed by any work of fiction.

small trees. Also our keen gardiners have dug flo[wer]
[fr]ont of each occupied barrack. We were however [gett]-
ing better soil in as most of our camp ground [is]
of sand in which nothing much will grow. Ne[xt]
the new plants are on the way it should look
[presen]table. We have just been working hard op[ening]
Xmas food parcels for this festive week, insi[de]
[were] several Xmas luxuries. Some have probably [gone]
to Nuutoskin at the Red cross centres. The parcels
[live] up to standard. Two or three days back a [letter]
from the Odell's; Alasdair apparently, has jo[ined and]
possesses Robert's great liking for high speed tra[ins and]
[r]oads. I have not played bridge recently, but [will be]
[at it] soon. The new five-suit game seem[s ...]

5
CODED CORRESPONDENCE WITH THE CAMPS

'Cryptography concerns communications that are deliberately designed to keep secrets from an enemy.'
(The Code Book. The Secret History of Codes and Code Breaking by Simon Singh)

It is clear from the previous chapter that it was critical to the success of getting maps and other escape aids into the camps, in order to assist the prisoners of war in planning and executing their escape, that regular contact with them was established and maintained by MI9. Under the Geneva Convention, prisoners of war were allowed to receive up to two letters and four cards monthly. MI9 took full advantage of this aspect of the Convention by establishing a system of coded letter exchanges with the camps. This was certainly the principal method of contact before wireless sets were smuggled into the camps.

The sheer volume and real significance of the coded letter traffic has begun to emerge, especially as a result of the successful deciphering of a cache of Lieutenant John Pryor's coded letters. Coded letters constituted a vital means for gathering intelligence from the prisoners of war and a channel through which they, in turn, could send their requests for escape and evasion maps of particular areas. The evidence assembled has thus substantially reinforced the significance of the coded letters which was highlighted in the author's discussion with Professor M. R. D. Foot in January 2012. There can now be no doubt that coded letters were a key part of the system which ensured that maps and other escape aids reached the camps in a timely fashion to aid the escapers in their attempts to return home.

The letters allowed the potential escapers to indicate precisely what they needed to support their escape plans and, in turn, alerted them not only to the nature of parcels being sent but also to the particular markings they would carry. The coded letters were of crucial significance to the way in which MI9 was able to ensure that information on escape aids, not the least of which were the maps, was communicated with the camps. Requests for, and the despatch of, items to aid the escapes which prisoners were planning were made through this important and covert channel of communication. Discovering how it worked, why, and just what it owed to the sheer ingenuity and commitment of both MI9 and the prisoners of war themselves provides another key part of the whole story of the escape and evasion mapping programme.

Innocuous letters to loved ones from British prisoners of war often contained hidden coded messages, such as this one John Pryor sent to his parents from Marlag and Milag Nord camp in December 1942. See page 140 for the full letter and page 141 for its decoding.

MI9'S CODING SYSTEM

In his fascinating book on the history of code use, Simon Singh highlights the importance of effective coded communication through history by monarchs and generals in governing their countries or commanding their armies and the fact that it was the threat of enemy interception of critical messages which was the catalyst in the development of codes and ciphers.

Singh, in particular, describes how the practice of utilizing codes has often had a dramatic impact on the course of history. How different, for example, might British history have been had Mary, Queen of Scots, not engaged in treasonable activity and incriminated herself through coded letters. The letters were intercepted and successfully deciphered by Queen Elizabeth I's Principal Secretary, Sir Francis Walsingham, generally acknowledged as the founder of Britain's Secret Intelligence Service. The letters between Mary and Anthony Babington did use an elaborate and very obviously enciphered text, and it was from Tudor times onwards that the art of enciphering and coding grew inexorably. To be strictly accurate, the word 'cipher' should be used in the context of MI9's system since it more appropriately describes the cryptographic process of hiding or disguising secret messages. Codes use words to disguise, for example, the identity of an agent or the nature of an operation: Airey Neave's codename in MI9 was Saturday and Overlord was the codename for the Allies' invasion of mainland Europe. However, the terms 'code' and 'coded' are used here as they are most commonly used and understood by laymen.

The system which MI9 elected to pursue, however, was, of necessity, rather different to the obvious encipher approach which would never have made it past the German censors. MI9 chose rather to develop a system of constructing letters which would not arouse suspicion by appearing quite innocuous, yet containing a hidden message. This was a distinctly alternative way of employing cipher alphabets. In essence, it did not matter if the general encryption method, or algorithm, became known; rather it was the specific key which identified the particular encryption which needed to remain secret. Even if the encoded message or cipher text was intercepted by the enemy, it could only be deciphered if the key was known. This was exemplified exactly in the deciphering of John Pryor's letters.

The general encryption method which MI9 employed has been known for some time since it had been publicized originally by Julius Green and later by Foot and Langley. The key comprised two Arabic numbers and an alphabet letter. In Green's case this was 5 6 O, and he showed how the code worked in his book, *From Colditz in Code*, published in 1971. The book was based entirely on his

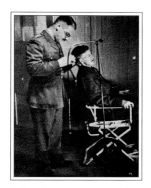

Julius Green, a code user in Marlag and Milag Nord and Colditz camps, working as a camp dentist.

personal experiences as a dentist in the field during World War II and specifically as a prisoner of war, eventually in Colditz. Green had been a Territorial Army reservist from his days at Edinburgh University in the early 1930s. He was commissioned in 152 Field Ambulance Brigade of the 51st Highland Division on 24 August 1939.

However, Pryor's individual key was not known and had to be identified in the course of this study. The numerical parts of his code were discovered early in the research as he had recalled it in his memoirs as 5 4, correctly, as it later transpired, but he had not recorded the alphabet letter. The discovery of the key was achieved by a lengthy process of elimination, testing each letter of the alphabet until a word which made sense was deciphered. This proved to be the letter S. Professor David McMullan, of the School of Mathematics and Physics at Plymouth University, provided considerable assistance in identifying the full key for John Pryor's code which enabled the deciphering of his letters. This greatly contributed to a more detailed understanding of the importance which the system of coded correspondence played in MI9's escape and evasion mapping programme. A full consideration of the construction of the code used in the letters is given later in this chapter.

SELECTION AND TRAINING OF THE CODE USERS

It is not surprising that coded correspondence between MI9 and the camps played such a key role in the escape and evasion mapping programme. One of the intelligence organizations that pushed hard for the creation of MI9 in 1939 was MI1. MI1 was the Military Intelligence Code and Cipher School which subsequently morphed into the Government Communications Headquarters, more generally known as GCHQ. It is, therefore, quite likely that MI9 staff also worked with MI1 in the development of the coded communication with the camps, although no confirmation of this likely connection has yet been discovered.

With the fall of France in June 1940, barely six months after the creation of the Branch, MI9 knew that over 50,000 Army personnel were in captivity and very few of them had been briefed on any aspect of escape and evasion, and none on coded correspondence. Indeed, at that stage they knew of only three prisoners of war who were code users, two in the Royal Air Force and one in the Royal Navy. They very quickly alerted the censors to the need to try to identify any letters from prisoners of war which they 'suspected of secondary meaning or of containing a private means of communication'. Through this means they were able to identify a few private codes which some individuals had had the foresight to establish with their families prior to deployment. MI9 contacted these individuals, initially using the private codes they had set up.

By early 1941, MI9 was in contact with men who had been briefed on the coded system prior to deployment and who were in turn briefing others in the oflags into the system. The big challenge was then to establish contact with the stalags. They managed this through padres, doctors and dentists who had the opportunity to minister to the needs of the other ranks in the stalags and could identify those most likely to be able to pick up the coded system and use it. They then briefed these men and slowly the system was spread through the camps. During 1941 the work grew to such an extent that a dedicated section had to be formed to cope with the increasing volume of coded letters being sent and received. The new section, Section Y, was formed in January 1942 and was attached to the Training School in Highgate. From these small beginnings, the coded system of contact with the camps grew and developed apace.

Potential coded letter writers were initially selected by the MI9 lecturers and trainers from officers who attended the training courses at the Training School or who they met during their lecturing visits to the various RAF camps and operational Army units. MI9 was on the lookout for men who appeared bright, responsible and discreet. These individuals were taken aside at the end of the lecture or training course and taught a simple code so that, if captured, they could conceal a coded message in an apparently routine letter to their family. Before the selected individuals were authorized as code users, they were required to do practice exercises to ensure that they were proficient enough with the system of encoding.

By September 1940 coded letters were being exchanged with men in prisoner of war camps in Germany and, at home, a scheme had been started to ensure that contact was maintained with the close family of any code user who had been captured. Local intermediaries were established on a regional basis and families were regularly briefed. MI9 was determined that their relationship with the families should be personal rather than being perceived as formal and official. The War Diary entry for August 1941 mentions the 'good relations and close co-operation' which had been established with the relatives and the extent to which it was increasing. Relatives clearly appreciated this and were known to have expressed their thanks to the War Office, indicating how they valued 'the War Office keeping in touch with them in a personal way' and not just behaving as an 'official machine'. It must have been some consolation to the families to

Naval prisoners of war at Marlag and Milag Nord, 1941. Julius Green is first left in the front row.

know that their sons were still doing their duty and continuing to contribute to the war effort whilst remaining in captivity. A letter headed MOST SECRET from MI9 on 16 December 1942 to Julius Green's mother told her that:

> For your private information, we are very glad to tell you that your son is continuing to do the most valuable work.

The seriousness of this work was then reinforced by the following message;

> Please do not show this letter to anyone outside the immediate family circle, and remember to burn our letter when read.

DEVELOPMENT OF THE CODES

From very early in 1940, the staff of MI9 were involved in devising codes. By April of that year, Lieutenant Commander Rhodes and Flight Lieutenant Evans were spending a great deal of their time in trying to perfect a suitable code. Initially known as the HK code, it had been suggested by Mr Hooker in the Foreign Office and was regarded by MI9 as the best idea put forward for a straightforward and simple code. The code was taught to the selected RAF and Army personnel. It was imperative that a detailed record of those briefed in the code was kept and that their families were also subsequently briefed, should their sons be unfortunate enough to be captured: as with all MI9's records, these were held on a detailed card index.

The system of codes steadily developed and related research work is regularly mentioned in the War Diary, although no detail about the construction of the codes and methods of encryption and deciphering was ever disclosed in the monthly reports. That detail was, however, included in other reports produced at the time and it is clear that work continued throughout the war on the development of the codes. There were eleven codes developed in all, some being retained for very specific use. Codes I, II, III and VI were apparently developed for use in general correspondence with the camps. Code III was developed during early 1942 and was in use in some of the camps later that year. Although regarded as an effective code, it apparently limited the length of the message when compared with Code II and the code users appeared to be more prone to making errors in their messages. While it had proved relatively easy to spot errors with Code II, it proved to be far more difficult to spot errors with Code III. By April 1942, the development of Code VI had progressed and was being taught to potential code users. Code V was specifically developed for use in Oflag IVC (Colditz). This code was established within the camp by communicating its detail through code users there who were already using Code II. The first message using the new code was received from Colditz in October 1942: it had been correctly encoded in every respect. Code IV was given to the Americans' MIS-X for their use. After the USA entered the war, MI9 had advised particular code users to brief selected US prisoners of war in the oflags, in anticipation of them being ultimately segregated. Code VII was used in the Middle East and Code VIII was retained for special operations, in particular those being organized by the Special Operations Executive (SOE). Codes IX, X and XI were apparently developed but had never been used by the time the war ended.

THE CODED LETTER WRITERS AND THEIR ROLE

Coded messages contained three different and important types of information: intelligence, the state of morale in the camps and planned escapes. The extent of the role played by the coded letters in terms of the intelligence which was actively sought and supplied is, perhaps, one of the more surprising aspects to emerge in the course of this study and one which appears previously to have been largely unacknowledged by historians. Certainly Julius Green's book, *From Colditz in Code,* revealed the extent of his own involvement in responding to MI9's requests for current intelligence information. By 1941 Green was in the Marlag and Milag Nord camp at Sandbostel, close to Bremen, and coincidentally the camp in which John Pryor was held at the same time.

The Marlag and Milag Nord camp at Sandbostel in 1941.

Green received a letter dated 11 February 1943, purporting to be from Philippa, possibly his sister, and ostensibly chatting about the mundane matters of the family's general health, especially that of 'Aunt Eleanor' and 'Daddy'. He was asked in the coded request:

> WE KNOW OF TWO LARGE OIL PLANTS WHICH WE CALL BLECHAMMER NORTH AND SOUTH ARE EITHER OR BOTH PARTLY WORKING OR COMPLETED. IS THERE A THIRD PLANT MAKING AVIATION PETROL NEAR HEYDERBRECK

Green replied on 24 March 1943 in a letter to his 'Dear Dad' which appeared to talk about war bonds and investments and which he signed 'Your affec. son Julie', his familial nickname. However, Green had managed to encrypt within his letter the response:

> BLECHAMMERS MAKING FROM 6 TO 9 THOUSAND QTS OIL DAILY I'VE ACCESS TO PLANT

As a dentist, Green was in the privileged position under the Geneva Convention of being allowed rather more freedom of movement than most prisoners of war, being a 'protected person' and regarded as a 'non-combatant'. Doctors and padres also belonged to this group and it is known that MI9 maximized the opportunities presented. In Green's case he felt on occasion overwhelmed by the amount of potentially valuable information he was able to acquire, commenting 'so much information was pouring in to me now that

I could not possibly encode it all'. Part of what he saw was evidence of the extermination programme of his fellow Jews which he described in his book in hauntingly moving terms:

> Once, on the way to visit the main camp at Lamsdorf, my guard and I had to change trains. In a siding we saw a train of closed trucks from which an intolerable stench issued . . . we walked back with the moaning and whimpering from the trucks in our ears . . . The trucks contained Hungarian Jews who were being transferred to an extermination camp for 'processing'. The sound has never left me and I still hear it.

It is not known whether Green ever reported this experience to MI9 or whether any action about the concentration camps was ever taken prior to liberation but it is clear that their existence was known to some of the prisoners of war.

Similarly, John Pryor, a captured naval lieutenant, was also providing intelligence which would clearly be of considerable assistance to the RAF for targeting purposes. In a letter he sent to his parents in December 1942, which at face value was describing the gardening activities in the camp and the Christmas parcels which had been received from the Red Cross, he encrypted a message which provided precise locational details about large ammunition dumps. The letter is shown later in this chapter while its detailed decoding is fully explained in Appendix 10.

John Pryor in 1951.

Since all Pryor's received correspondence was lost during his repatriation in 1945, there was initially no indication that he was responding to a received request or simply providing acquired intelligence. However, it subsequently emerged that he was, in effect (just like Julius Green), responding to requests for information, as copies of some coded letters to the camps were kept by MI9.

MI9 also used coded letters to assess the morale of prisoners of war in the many camps with which they were able to establish and maintain contact. In March 1941 Pryor had written a letter to his parents mentioning the vegetable garden, receipt of parcels and a letter from home. In it he had managed to encrypt the coded message:

> OWING SUPPOSED CONDITION PRISONERS CANADA WE MAY SUFFER HERE URGENT INQUIRIES NECESSARY

These words conveyed information that there were fears amongst the prisoners who apparently expected the German guards would take retaliatory

Prisoners at Stalag Luft III (Sagan) tending their garden. John Pryor frequently wrote about the challenges of maintaining the garden in the sandy conditions of Marlag and Milag Nord camp.

action over the reported mistreatment of German prisoners of war being held in Canada.

DECODING REQUESTS FOR ESCAPE AIDS

The principal interest of the coded traffic, however, in so far as this study is concerned, is to demonstrate the importance of contact in requesting and providing the wherewithal to assist planned escapes. In a letter sent in May 1942 to his parents, which at face value purported to talk about the books he wanted and the extent to which the camp gardens were improving, Pryor managed to encrypt the following message:

> CLOTHING AND LOCAL MAPS OBTAINED REQUIRE SOME OF BORDERS
> ESPECIALLY SWISS PASPORT INFORMATION AND RENTEN MARKS

Despite the spelling mistake in 'passport', it was unmistakeably a request for assistance to support an escape. Since this letter was the first to be decoded in this study, it provides an ideal example of the methodology employed to uncover the hidden message, shown later in this chapter. There is no indication as to whether he received a response to his request for maps of the Swiss border; however, it is more than likely that he did, since the Swiss border was certainly an escape route recommended by MI9.

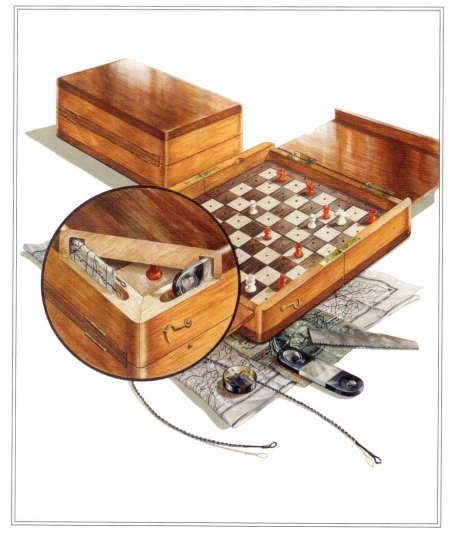

An MI9 drawing of a chess set and its secret compartments, similar to the set that John Pryor described.

However, it is known from Pryor's own memoirs written in his later life that he did receive early in 1943 a coded message in a letter from his (unknown) Auntie Florrie alerting him to expect a parcel from the 'London Victuallers' Company' (almost certainly the Licensed Victuallers' Sports Association, a known cover organization for MI9). He recounted how he was then detailed by the Escape Committee to be present in the Post Room to ensure that the parcel was not opened by the German censors, but was appropriated by the prisoners of war and carefully whisked away to be hidden until it could be opened and explored in more secure surroundings. He continued by describing the chess set he found inside the parcel and the subsequent splitting open of the chessboard to reveal its hidden contents, namely German currency, hacksaw blades, copies of a German *Ausweis* (work and travel pass or permit) and examples of letters of introduction. The items were handed over to the Escape Committee. Interestingly, when it came to Pryor's own escape (described in detail in Chapter 7), the route he attempted was not across the Swiss border but via the Baltic ports which were located geographically very much closer to the Marlag and Milag Nord camp.

There is also evidence contained in a letter sent in June 1943 to Lieutenant F. C. Hamel of the Royal Naval Volunteer Reserve (RNVR), also a prisoner in the Marlag and Milag Nord camp, purporting to be from his mother. It carried, however, a hidden message from MI9 which read;

> DECCA RECORDS ALGER SOAK COVERS TO GET MAPS

It appeared to alert him to the despatch of a parcel to an individual named Alger. The message also shows that MI9 was continuing to use gramophone records but, instead of hiding the maps in a laminated compartment in the centre of the record, they were rather encapsulating them in the record covers, probably in similar fashion to the method used to hide them inside playing cards.

VOLUME AND PASSAGE OF CODED LETTER TRAFFIC

MI9 responded promptly to requests for information and equipment to aid escape, such as currency, passes, maps, saws and compasses. They could advise on routes, specifically which were the best and most likely to result in a successful escape and which to avoid. By 28 February 1941 most of the coded letters received in London contained requests for escape materials and German currency. By this time, MI9 had developed Code II and were also briefing men involved in particular covert operations. One disadvantage of the coded letter was the length of time it could take to reach its destination in either direction. It often took weeks and sometimes months for the letters to arrive. It appears that MI9 enjoyed more success in maintaining contact with the camps in Germany than with those in Italy. The relative lack of success in Italy was ascribed to the inefficiency of the Italians as 'mail and parcels could take over a year and were often lost or destroyed'. The transit and delivery of letters and parcels to the camps in Germany appeared to be rather more swift and efficient. It is an irony that Teutonic efficiency apparently contributed to the success of the escape and evasion programme.

English and French prisoners of war receiving mail at a prison camp in the Castle of Wülzburg, Bavaria.

The numbers of letters steadily increased as more personnel were briefed in their use and as more men were captured. In the month of March 1941, sixty coded letters were received in MI9 and fifty-six were despatched. Four months later the total number of acknowledgement slips for parcels and coded letters received from the camps had increased to 4,279 or approximately 138 every day. Incoming letters were picked up by the Censors who were given a list, known as the Special Watch List, by MI9 of the names to look for and ensure that any letters from them were redirected to MI9. By December 1941, there were 928 names on the list and MI9 knew it could potentially rise very quickly to 1,500. They recognized how challenging and taxing it was for the Censors to pick up all the coded letters. This identification of which letters to decode was, after all 'the vital link in the chain and, if broken, the whole structure will collapse'. MI9, therefore, made every effort to ensure that the list of names was kept as current as was humanly possible in order to make the Censors' job less onerous. The decoding work was regarded as urgent and always treated as a priority by MI9. Once they had decoded the hidden messages, the letters were passed on to the families.

Matters relating to escapes were dealt with by MI9; anything relating to technical or operational intelligence was passed immediately to the appropriate departments of the three Services. It was not unknown at this stage of the war for over a dozen intelligence reports to be received in a month. They were also succeeding in speeding up the transit of coded letters so that, by the end of June 1941, the War Diary entry notes an example of a letter that was despatched by MI9 on 10 May reached the particular camp during the first week in June and a reply was received in MI9 on 29 June. By this time there were seventy-two coded correspondents operating in the camps in Germany and Italy.

Prisoners of war who were being transferred to different camps would often pass on their code to a fellow prisoner to ensure that contact with the camp was not lost. It was also the case that some prisoners who were to be moved were briefed by code holders so that the code could operate from the camp into which they were being moved. Such was the case with John Pryor who indicated in his memoirs that it was during January 1941 while he was in Oflag VIIC/Z at Titmoning that he was approached by a junior Army officer who had heard that the naval prisoners were to be moved to another camp. He asked Pryor if he would be prepared to learn a secret code so that messages could be hidden in his normal letters home. Pryor confessed that he was initially suspicious, thinking that the officer might be a German plant. He approached the Senior British Officer, General Fortune, and was reassured that the approach was genuine. Pryor proceeded to learn the code.

This practice of teaching the code methodology to others indicates the close cooperation which existed between the prisoners of different Services and was a practice which Pryor himself used later on in his captivity. In March 1942 he wrote a letter in which he hid a message indicating that he had taught codes to four others, namely:

ELDER 6 5 M HEAP 5 7 K HAMEL 7 5 O WELLS 5 6 J

There is an indication in his letter that he had been asked to do this since the hidden message starts by indicating that he had understood the coded letters dated 24 December and 3 January which he had received. Interestingly, but inexplicably, he found it necessary to resend the four codes he had allocated by repeating the message in a letter he sent two weeks later in early April. Not surprisingly, the two letters are superficially about quite different topics and the hidden messages, as a result, are encoded differently though relaying the identical information: in itself, this demonstrates the sheer versatility of the coded letter system.

It is clear from Pryor's memoirs just how much effort went into the planning and execution of escapes. He was able to use the MI9 code and encrypt messages in some twenty-one letters to his parents which contained both intelligence and requests for assistance. His efforts serve as a stark reminder of the *raison d'être* of MI9 and the extent to which prisoners of war had taken on board the escape philosophy. Of necessity the codes were hardly simple; they had to avoid detection by the German censors. To encode letters, as Pryor did, must have taken many hours of intellectual effort in circumstances which were hardly conducive to such activity. It is to his eternal credit, and that of the many other prisoners of war, that they continued to do exactly what MI9 asked them to do, harass the Germans at every opportunity, attempt to escape, be the proverbial thorn in the enemy's side, and provide intelligence in response to requests. In return, MI9 stuck determinedly to their part of the 'agreement', providing all the escape aids in their considerable armoury to aid the planned escapes. They also ensured that all families who had sons in captivity and who had been schooled in the art of sending and receiving coded letters knew exactly what was happening.

By July 1941, MI9 was operating a network of 254 code correspondents, which rose to 928 by December. By February 1942 they were in contact with twenty-seven camps in Germany and Italy and estimated that the system allowed them to maintain contact with over 62,000 prisoners of war. They also knew of 465 attempted escapes by that time.

The value of the coded letters as a source of potentially useful intelligence information was being steadily realized by other Government departments. They increasingly passed questions to MI9 to be directed through the coded letter system to prisoners of war in particular camps where they were known to work outside the camps alongside Germans and could report on what was going on in the surrounding area. The occurrence of this practice is suggested by the volume of intelligence passed by Julius Green.

There was one reported compromise of the system by British government officials and this occurred when the British Military Attaché (MA) in Stockholm acted unilaterally in sending parcels to prisoners of war and included instructions on how to use one of the codes. Three camps, Stalag Luft III (Sagan), Oflag IX and Oflag VIB, were able to warn MI9 what the MA had done, that it had been discovered by the Germans and the whole system was potentially compromised. Major Winterbotham, an MI9 officer, was then immediately despatched on 7 May 1942 to Stockholm to order the MA to stop sending parcels. It was never clear why he had taken such action in the first place but the extent to which his action potentially compromised the whole system was clear. MI9 immediately stopped using Code III and speeded up the introduction of the Code VI. Code users were able to alert MI9 on more than one occasion to problems such as the discovery of escape aids in parcels. Interestingly, no reported discovery of maps has been found.

THE CONSTRUCTION OF CODED LETTERS

How, then did the codes work and specifically how were the letters constructed with the messages hidden? Coded letters were immediately identifiable by the form of the date: 22/3/42, rather than the more usual form of 22nd March 1942; this numeric date form indicated that the letter contained a coded message. Underlining the signature also confirmed that the letter carried a coded message, as did the inclusion of the word 'very' in the closing line, e.g. 'yours very sincerely'.

The length of the message was determined by multiplying together the number of letters in the first two words of the first line of the letter after the salutation. If the first two words each comprised five letters, then the message was 5 × 5 = 25 words in length and a grid 5 × 5 needed to be constructed. If the words had different numbers of letters in them, then the first word indicated the number of words running horizontally in the grid and the second word indicated the number of words running vertically in the grid.

Using the example of a letter starting with two five-letter words, a square grid was drawn with 25 squares in it, i.e. 5 × 5. The rest of the first line of the letter was ignored.

Moving to the second line of the original letter, it was necessary to use the numerical part of the personal key of the individual coder. Julius Green's key was 5 6 O, so the fifth and sixth words alternately were selected from the letter and placed in the grid, starting in the top left hand corner and working left to right on each consecutive line as indicated in the grid below.

1	2	3	4	5
6	7	8	9	10
11	12	13	14	15
16	17	18	19	20
21	22	23	24	25

When the grid was complete, however, the message was read starting bottom right, reading across and then diagonally, across and diagonally, etc., as shown over the page, reading in sequence 1 to 25.

25	24	22	19	15
23	21	18	14	10
20	17	13	9	6
16	12	8	5	3
11	7	4	2	1

The last word of the message, therefore, went into the first box on the grid and the first word of the message went into the last box of the grid. Different code users were briefed to use different word counts, enshrined in the numerical part of their personal code. That of Julius Green was the fifth word followed by the sixth word, whereas that of John Pryor was the fifth word followed by the fourth word. It is clear that there were variations in the codes given to each code user and MI9 needed to have kept a very detailed record on their card index system.

When the word 'but' occurred in the sequence, i.e. the fifth or sixth word in Green's key, it signified the end of the message which should, of course, also have accorded with the completion of the grid. Words which were hyphenated or had an apostrophe in them were to be treated as one word for the purposes of counting, e.g. 'two-days', 'I've'.

When the word 'the' occurred in the sequence, i.e. the fifth or sixth word in Green's key, this indicated that the alphabet code started. Using Green's key of 5 6 O, his alphabet code consisted of the alphabet starting with O. The alphabet grid was written down in three columns, as shown below, and after Z, a full stop occurred. This gave twenty-seven items and three columns of nine letters, each with an associated three-digit number.

O 111	P 211	Q 311
R 112	S 212	T 312
U 113	V 213	W 313
X 121	Y 221	Z 321
. 122	A 222	B 322
C 123	D 223	E 323
F 131	G 231	H 331
I 132	J 232	K 332
L 133	M 233	N 333

Commencing with the next sentence after the occurrence of 'the', the first letter of each consecutive word was written down in groups of three. Each letter was then identified on the alphabet grid shown above and then the number of the column (1, 2 or 3) in which the letter occurred was noted down. Thus, if the sentence started 'Which brings the' the significant letters were W B T. Each group of three letters was used to signify one letter in the table. For W B T, using the alphabet grid above, W was in column 3, B was in column 3 and T was also in column 3. This represented the number 333 which, using the table again, indicated that the letter required was N.

A word was built up by continuing to take the first letter of consecutive groups of three words until the full stop occurred in the coding (i.e. 122 in the above example). The decoding then continued by moving to the next sentence and reverting to the word count. Again, there were individual variations since it could involve moving to the next paragraph and then reverting to the word count.

In the same way that there were variations in the word count, it was proved by Professor McMullan that there were also variations in the alphabet code. He identified the letter S as the missing alphabet key in John Pryor's code by a lengthy elimination process of all other letters of the alphabet. The S alphabet code is shown below.

S 111	T 211	U 311
V 112	W 212	X 312
Y 113	Z 213	. 313
A 121	B 221	C 321
D 122	E 222	F 322
G 123	H 223	I 323
J 131	K 231	L 331
M 132	N 232	O 332
P 133	Q 233	R 333

Whichever alphabet code was used, it was based directly on a form of modular arithmetic where the pattern of three is key to constructing the coded message and deciphering it.

While MI9 did initially describe their codes as 'simple', it is clear that, from a layman's perspective, they were far from simple and, in order to avoid the searching oversight and scrutiny of German censors, it is hardly surprising. The code user was essentially required to reverse engineer his message. He started with the message to be encoded, constructed the grid, numbered it, placed those numbers/words in the correct order in his letter with the requisite number of spaces alternating between, reflecting his own personal numerical code, and then sought to construct a sensible sounding but totally innocuous letter linking it all together. The whole exercise required considerable intellectual application, time and commitment to plan and construct the message ahead of time. Coders were then in a position to maximize the use of the limited opportunity they were given by their German captors to write letters home, and commit that opportunity to a covert purpose.

It must have been a considerable morale booster for the prisoners of war to know that, despite their loss of liberty, they were still in a position to contribute in a very significant way to the war effort by providing valuable intelligence information, and also to have the reassurance of knowing that an entire organization was working at home to aid their attempts to escape and to ensure that their families were being kept briefed about their continuing contribution. In addition, it would have relieved their considerable boredom during long months and years spent in captivity with ample 'leisure' time on their hands.

Three British prisoners of war produce news sheets during their 'leisure time' at Stalag Luft III.

Staff at the Army Post Office 'somewhere in the Midlands', 1944. Letters sent to prisoners in the camps were sent through here.

DECODING A HIDDEN MESSAGE

As an example, over the page there is the text of a letter dated 7/5/42 which John Pryor wrote home. The form of the date, the underlined signature and the inclusion of the word 'very' in the last line of the letter indicate that this is a coded letter. The exercise to decode the letter is marked up on the accompanying version, where significant words and letters are highlighted in red. The following method of decoding was employed to discover the hidden message.

CODED CORRESPONDENCE WITH THE CAMPS   133

John Pryor's coded letter

7/5/42

My Dear Mummy & Daddy, Last week I received a short letter from Robert. The envelope had the marks of five of the RAF censors. I can't imagine what his new number on the envelope means, maybe he has been turned over to rather different occupations, which of course I can't know anything about. I am glad the information I sent you, especially about the Uffa Fox and other books of the sailing variety, reached you. As regards other possible books, my present desires seem mostly for interesting literature of events in our country's history. A subject I am unfortunately very weak in.

The gardens are improving, borders of wire which require a net to keep the football off. Possibly a move to Marlag Nord in the late summer may prevent getting all advantages from some of the later plants, but we hope not. During the course of the last days we obtained some chairs, from local sources I believe, also some tables, which give the recreation room a much better appearance. A few weeks ago we arranged a rather useful scheme, so men could get "legergeld" by remitting money at home for that received here. After another year – year and a half my clothing requirements etc. will be but a mere trifle.

As I have received all your clothing parcels sent off up to the end of 1941. The suitcase in the last one arrived a bit battered but still quite usable. Private cigarette and tobacco parcels seem to becoming in better now, so I am hoping to get another shortly as my last one arrived in November. Now I must stop, hoping you are all as well as I am. Remember me especially to Marj. Your very loving son.

John.

John Pryor's coded letter with hidden message highlighted

7/5/42

*My Dear Mummy & Daddy, Last week **(4x4 grid)** I received a short letter from Robert. The envelope had the **marks (5)** of five of **the (4 - start alphabet code)** RAF censors. **I c**an't **i**magine **w**hat **h**is **n**ew **n**umber **o**n **t**he **e**nvelope **m**eans, **m**aybe **h**e **h**as **b**een **t**urned **o**ver **t**o **r**ather **d**ifferent **o**ccupations, which of course I can't know anything about. I am glad the **information (5)** I sent you, **especially (4)** about the Uffa Fox **and (5)** other books of **the (4 – start alphabet code)** sailing variety, reached you. **A**s **r**egards **o**ther **p**ossible **b**ooks, **m**y **p**resent **d**esires **s**eem **m**ostly **f**or **i**nteresting **l**iterature **o**f **e**vents **i**n **o**ur **c**ountry's **h**istory. **A** subject **I a**m unfortunately very weak in.*

*The gardens are improving, **borders (5)** of wire which **require (4)** a net to keep **the (5 – start alphabet code)** football off. **P**ossibly **a m**ove to **M**arlag **N**ord **i**n **t**he **l**ate **s**ummer **m**ay **p**revent **g**etting **a**ll **a**dvantages **f**rom **s**ome **o**f the later plants, but we hope not. During the course **of (4)** the last days we **obtained (5)** some chairs, from **local (4)** sources I believe, also **some (5)** tables, which give **the (4 – start alphabet code)** recreation room a much better appearance. **A f**ew **w**eeks **a**go **w**e **a**rranged **a r**ather **u**seful **s**cheme, **s**o **m**en **c**ould **g**et "legergeld" by remitting money at home for that received here. After another year – year **and (5)** a half my **clothing (4)** requirements etc. will be **but (5 – indicates end of message)** a mere trifle.*

As I have received all your clothing parcels sent off up to the end of 1941. The suitcase in the last one arrived a bit battered but still quite usable. Private cigarette and tobacco parcels seem to be coming in better now, so I am hoping to get another shortly as my last one arrived in November. Now I must stop, hoping you are all as well as I am. Remember me especially to Marj. Your very loving son.

John.

The first two words after the salutation are 'Last week': a 4 × 4 grid is, therefore, constructed. Moving to the second line of the letter and using Pryor's numerical code of 5 and 4, the fifth word is 'marks', so this goes into the top left box of the grid and is the final word of the message.

marks			

The fourth word after this is 'the', which indicates that the alphabet code starts at this point. Pryor's alphabet code started at S and is shown below.

S 111	T 211	U 311
V 112	W 212	X 312
Y 113	Z 213	. 313
A 121	B 221	C 321
D 122	E 222	F 322
G 123	H 223	I 323
J 131	K 231	L 331
M 132	N 232	O 332
P 133	Q 233	R 333

Beginning with the next sentence, the first letter of each consecutive word is written down in groups of three. Thus, the decoder moves to the sentence which starts 'I can't imagine' and writes down:

I C I

I is in column 3 in the table above, as is C, so:

I C I = 333

In the grid the number 333 represents R, so the decoder records:

I C I = 333 = R

The decoder lists the first letter of every word in groups of three, which gives the following:

I C I = 333 = R
W H N = 222 = E
N O T = 232 = N
E M M = 211 = T

136   GREAT ESCAPES

H H B = 222 = E
T O T = 232 = N
R D O = 313 = .

The letters spell 'renten', the name of the currency in Nazi Germany, which becomes the second word on the grid.

At the point where the full stop occurs, the decoder reverts to the 5 4 sequence at the start of the next sentence in the letter while maintaining the correct alternating rhythm. Having finished on the fourth word at the previous stage, this time the fifth word is counted. This gives the word 'information' which becomes the third word on the grid. The following fourth word is 'especially' which becomes the fourth word on the grid. The fifth word is 'and' which becomes the fifth word on the grid.

marks	*renten*	*information*	*especially*
and			

The next fourth word is 'the', signalling that the alphabet code starts again at the beginning of the next sentence. Taking the first letter of each word and setting them out in groups of three produces the following:

A R O = 133 = P
P B M = 121 = A
P D S = 111 = S
M F I = 133 = P
L O E = 332 = O
I O C = 333 = R
H A S = 211 = T
I A U = 313 = .

It contains a spelling error but is nonetheless recognizable as the word 'passport', which becomes the sixth word on the grid.

Moving to the start of the next complete sentence in the letter and picking up the 5 4 rhythm, the fifth word is 'borders' which becomes the seventh word on the grid. The following fourth word is 'require' which becomes the eighth word on the grid. The following fifth word in the letter is 'the' which indicates that the alphabet code starts again, as follows over the page:

P A M = 111 = S
T M N = 212 = W
I T L = 323 = I
S M P = 111 = S
G A A = 111 = S
F S O = 313 = .

The ninth word on the grid is 'swiss'.

Moving to the start of the next complete sentence in the letter and keeping the 5 4 rhythm, the fourth word is 'of' which becomes the tenth word on the grid. The following fifth word is 'obtained' which becomes the eleventh word on the grid. The next fourth word is 'local' which becomes the twelfth word on the grid. The following fifth word is 'some' which becomes the thirteenth word on the grid.

marks	renten	information	especially
and	pasport	borders	require
swiss	of	obtained	local
some			

The following fourth word is 'the', which indicates that the alphabet code starts again at the beginning of the next sentence. This produces the following:

A F W = 132 = M
A W A = 121 = A
A R U = 133 = P
S S M = 111 = S
C G L = 313 = .

The fourteenth word on the grid is 'maps'.

Moving to the start of the next complete sentence in the letter and keeping the 5 4 rhythm, the fifth word is 'and' which becomes the fifteenth word on the grid. It is worth noting at this point the deliberate repetition of the word 'year' which makes no sense in English but is clearly designed to ensure that the fifth word is 'and' so that the hidden message will make sense. The following fourth word is 'clothing' which becomes the sixteenth and final word on the grid. To reinforce the point, the fifth word after this point is 'but' which confirms the end of the coded message: the introduction of the apparently superfluous 'etc.' ensured that 'but' occurred as the fifth rather than the fourth word. Placing all these words in their correct consecutive position on the grid produces the following:

marks	*renten*	*information*	*especially*
and	*pasport*	*borders*	*require*
swiss	*of*	*obtained*	*local*
some	*maps*	*and*	*clothing*

Starting in the bottom right corner and reading across and diagonally in sequence, the message reads:

> CLOTHING AND LOCAL MAPS OBTAINED REQUIRE SOME OF BORDERS ESPECIALLY SWISS PASPORT INFORMATION AND RENTEN MARKS

Leaving aside the misspelling of 'passport', it is a clear request for maps of the Swiss border, and information which would allow them to produce passports. It also carries an indication that they have already been able to obtain local maps and clothing. At the time of this letter (May 1942) Pryor was at the Westertimke camp near Sandbostel in northern Germany. Asking for maps of the Swiss border is perhaps surprising, since the distance that would need to be covered was considerable. Escape via the Baltic ports, which is what Pryor and others subsequently attempted (see Chapter 7), would have been more likely. The message does, however, serve as a stark reminder of the *raison d'être* of MI9 and the extent to which prisoners of war had taken on board the escape philosophy.

While the decoded letter shown above was concerned with planning escapes, Pryor also passed intelligence information back to MI9, as shown by the following letter, sent in December 1942. His original letter is shown first. The Censors would have been required to identify this as a coded letter, using such signs as the form of the date and the underlined signature.

Numerical writing style for the date and an underlined signature were both signs that a letter contained a coded message. See overleaf for the full letter.

22/12/42.

My dear Mummy & Daddy, The camp's appearance is looking quite smart now as the main road and paths inside the wire have been lined with small trees. Also our keen gardeners have dug flower beds in front of each occupied barrack. We were however forced to bring better soil in as most of our camp ground boasts only of sand in which nothing much will grow. Next spring when the new plants are on the way it should look quite respectable. We have just been working hard opening up our Xmas food parcels for this festive week, inside they contain several Xmas luxuries. Some have probably been on view to Ruth&John at the Red cross centres. The parcels are certainly up to standard. Two or three days back a letter came from the Odell's; Alasdair apparently, has joined up and possesses Robert's great liking for high speed travel on the roads. I have not played bridge recently, but hope of a rubber soon. The new five-suit game sounds the most complicated affair. My small model is really well underway and shows quite definite signs by now of resembling a real whaler. The contents of rubbish dumps etc. are really just the thing for getting the odd little bits of wood and tin for it! As regards the clothing parcel suggestions. There is nothing I need really. But you already understand the few odd consumable things that one needs. I shall really be quite content even if they are underweight. Recently we had a large number of books, but mine have not arrived yet. Heaps & heaps of love
your loving son
John.

John Pryor's letter sent from Marlag and Milag Nord camp in December 1942. It contains hidden intelligence information.

John Pryor's coded letter with hidden message highlighted

<u>22/12/42</u>

*My dear Mummy and Daddy, The camps **(3 x 5 grid)** appearance is looking quite smart now as the main **road (5)** and paths inside **the (4 – start alphabetical code)** wire have been lined with small trees. **A**lso **o**ur **k**een **g**ardeners have **d**ug **f**lowerbeds **i**n **f**ront **of e**ach **o**ccupied **b**arrack. **W**e **w**ere **h**owever **f**orced **t**o **b**ring **b**etter **s**oil **i**n **a**s **m**ost **of o**ur **c**amp **g**round **b**oasts **o**nly **of s**and **i**n which nothing much will grow. Next spring when the **new (5)** plants are on **the (4 – start alphabet code)** way it should look quite respectable. **W**e **h**ave **j**ust **b**een **w**orking **h**ard **o**pening **u**p **o**ur **X**mas **f**ood **p**arcels **f**or **t**his **f**estive **w**eek, **i**nside **t**hey **c**ontain **s**everal **X**mas luxuries. Some have probably been **on (5)** view to next-of-kin **at (4)** the Red cross centres. **The (5 – start alphabet code)** parcels are certainly up to standard. **T**wo **o**r **t**hree **d**ays **b**ack **a l**etter **c**ame **f**rom **t**he **O**dell's; **A**lasdair **a**pparently **h**as **j**oined **u**p **a**nd **p**ossesses **R**obert's **g**reat **l**iking for high speed travel on the roads. I have not played **bridge (5)** recently, but hope **of (4)** a rubber soon. The **new (5)** five-suit game sounds **the (4 – start alphabet code)** most complicated affair. **M**y **s**mall **m**odel **i**s **r**eally **w**ell **u**nderway **a**nd **s**hows **q**uite **d**efinite **s**igns **b**y **n**ow **of r**esembling **a r**eal whaler. The contents of rubbish **dumps (5)** etc. are really **just (4)** the thing for getting **the (5 – start alphabet code)** odd little bits of wood and tin for it! **A**s regards the **c**lothing **p**arcel **s**uggestions. **T**here **i**s **n**othing **I n**eed really. **B**ut **y**ou ^**a**lready^ **u**nderstand **t**he **f**ew **o**dd **c**onsumable **t**hings **t**hat **o**ne **n**eeds. **I** shall **r**eally be quite content even if they are underweight. Recently we had a **large (5)** number of books, **but (4 – indicates end of message)** mine have not arrived yet.*

*Heaps & heaps of love
your loving son
<u>John.</u>*

The text with the hidden message highlighted is shown above, while the full explanation of the decoding is given in Appendix 10. The message hidden within this letter about the camp gardens and Christmas preparations was:

LARGE MUNITION DUMPS JUST SOUTH OF NEW BRIDGE AT NARKAU ON NEW BERLIN MARIENBURG ROAD

ORGANIZATION IN THE CAMPS

The level of organization in the camps dedicated to planning and executing escapes has been described in Chapter 4. Such detailed organization was also maintained so far as the coded letter system was concerned. The ingenuity required to compose a message that seemed like an innocent and innocuous letter to the family back home must have been very great and would have needed some considerable time. Time was something the prisoners of war had in abundance but they would also have needed their colleagues to keep watch to ensure that the activity was not discovered by the guards. The SBOs and the Escape Committees decided which messages were sent and by whom. They alone knew the identities of all the coded letter writers in each camp. New arrivals were interrogated by them and, if they knew the code, they were ordered not to send messages independently but only when told to do so.

There were reports from at least one camp, Stalag Luft III (Sagan), at the end of the war that mistakes had been made by MI9. The criticisms mentioned use of the same typewriter for letters to different individuals in the same camp; different signatures by the same fictitious individuals; one prisoner's code key

A watchtower at Stalag Luft III. The official camp history criticized some mistakes made by MI9 in its communications with the camp.

142 GREAT ESCAPES

marked on the envelope instead of his prisoner of war number; use of similar notepaper for letters to different code users; and a lack of understanding in MI9 of the difficulties of keeping detailed records in the camps of all coded messages sent. Such detailed criticisms could only have been made by a team that was monitoring each and every letter. While the prisoners of war in each camp apparently knew that some means of secret communication between the camp and the UK existed, its precise form was known only to the SBOs, the Escape Committees and the code users. The activity was certainly managed in a highly controlled and meticulous fashion with considerable attention to the detail and the security.

ASSESSING THE CODED LETTER PROGRAMME

Two particularly remarkable points emerge from studying the exchange of coded letters. The first is the considerable degree of danger to which prisoners of war exposed themselves, not simply in planning their own escapes but also in sending intelligence information back to MI9. While the penalty for attempted escape was thirty days in solitary confinement, the penalty for providing intelligence was likely to be a session with the Gestapo for alleged espionage activities. Secondly, the sheer cleverness of the codes is strikingly impressive. Although the codes employed were described as simple by MI9, the finesse of the methodology was such that, as each was unique to the user, it was unlikely to be discovered. Since each code comprised three elements, two Arabic numbers and a letter of the alphabet, there were 2,600 possible permutations. This allowed for the existence of 2,600 coded letter writers, if each was to use a unique code. By December 1941 there were already 928 coded correspondents in operation and it was anticipated that this number would likely increase to 1,500 very rapidly. By the end of the war in 1945, a total of 12,500 coded letters had been despatched and received by MI9. The peak year was 1943, with a total of almost 4,500 in that year alone. By 1944, radio communications had been established in many of the camps and, together with the collapse of Italy, this resulted in a reduction of the coded letter traffic.

There is no evidence that a coded letter was ever deciphered by the Germans (or the Italians) or, indeed, that they were even aware of their existence, until the action by the British Military Attaché in Stockholm which might well have wrecked the entire system, but apparently did not. The difference in national approaches to the use of codes is also relevant. The Germans used a technical approach with machines to encrypt their coded messages, of which ENIGMA is the most famous. The British had always tended to regard codes and encryption as an intellectual exercise, preferring to use people rather than machines. Crosswords and chess were well-known as leisure activities in the UK; it was essentially the same approach of the application and discipline of intellectual thought and logic which allowed the coded letter approach to succeed as effectively as it evidently did. It was arguably this significant difference in national traits and practices which allowed MI9 to succeed with this most important link in the whole escape chain: it allowed the prisoners of war to indicate what they needed by way of maps and other escape aids and ensured that MI9 could respond by reporting what was being sent, to whom and when.

German soldiers in the field enciphering a message using the ENIGMA coding machine.

MI9 regarded the whole operation of coded correspondence as a success story. From their perspective, they took inordinate care with the operation of the entire system. They also recognized that integral to the success had been the outstanding security and planning which had taken place in the camps. It is, therefore, perhaps one of the most surprising aspects of the whole story of the coded letters to realize the extent to which it has not been addressed in detail by historians to date. Certainly Foot and Langley mentioned it in the history of MI9 which was published over thirty years ago and Green's book tracing his own role as a coded letter correspondent with MI9 was published over forty years ago. Foot and Langley acknowledged Green's book as their primary source and Foot emphasized in conversation as recently as January 2012 the importance of the coded letters. One might reasonably ask why that importance was not recognized by other historians, especially those who more recently have had access to rather more open MI9 related files than Foot and Langley ever did in the 1970s. One might conjecture that the story has perhaps been overshadowed by that of Bletchley Park, ENIGMA, Alan Turing and his staff. For whatever reason, it is certainly the case that the coded letter story appears to have engendered no real interest or consideration by historians and, more particularly, by those involved in the history of British intelligence in the twentieth century. And yet, when one considers precisely what MI9 managed to do and what the prisoners of war managed to contribute in terms of current intelligence as well as planning escapes, it has to be acknowledged that this was a quite stunning contribution to the war effort.

The construction of the network of coded correspondence bears testimony to the extent to which escape had become a professional undertaking involving painstaking planning and execution. The programme of coded correspondence was every bit as carefully planned and executed in all its aspects as the map production programme and was a vital link in the escape chain. By this stage of the unfolding story, it is very clear that MI9 demonstrated the same professionalism, attention to detail and application in every aspect of the escape and evasion programme on which they had embarked.

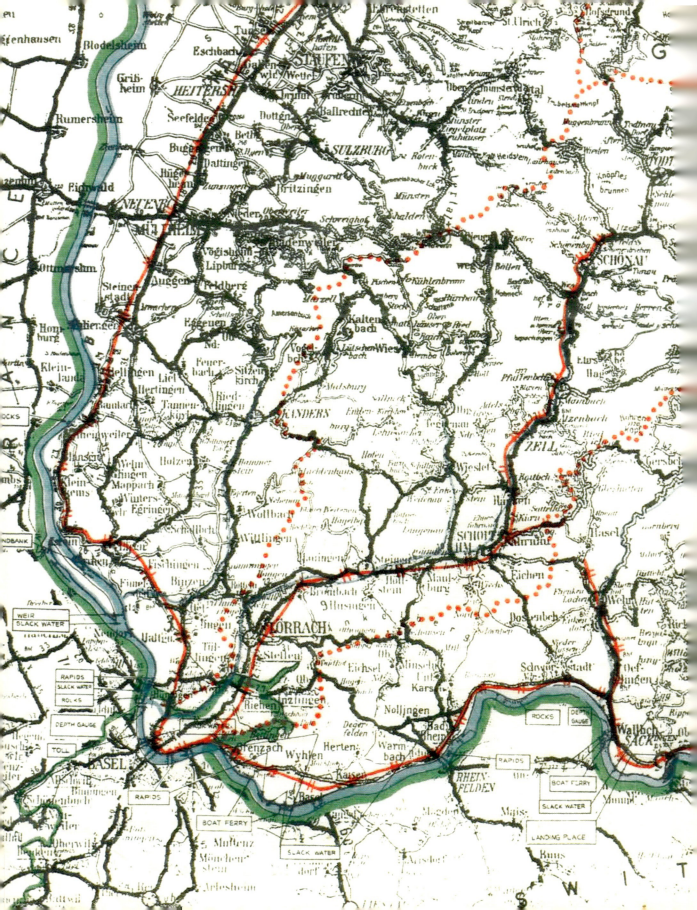

6

THE SCHAFFHAUSEN SALIENT AND AIREY NEAVE'S ESCAPE

'Life's battles do not always go to the stronger and faster man, but sooner or later the man who wins is the man who thinks he can.'
(Anonymous, found in Airey Neave's private papers)

The way in which the mapping programme was conceived, the significant resources which were dedicated to it, the ways of getting the maps into the camps and the method of coded contact have been set out in earlier chapters. There remains, however, the whole issue of how the maps were used in actual escapes and the extent to which they successfully, or otherwise, fulfilled their role. This aspect is first explored through looking at the escape route through Switzerland and Airey Neave's escape using it. Neave's escape, and further examples that follow in the next chapter, describe how the maps were used and the extent to which they were a key aspect in escape attempts: in essence, they demonstrate the geography of escape.

MI9'S ESCAPE ROUTES

Before considering individual escape stories, it is useful to appreciate some important aspects of the escape routes which were selected by MI9 and the critical extent to which MI9's rationale for their choices was reflected in the escape and evasion mapping programme. It is certainly the case that there were more attempts and more successful escapes from German prisoner of war camps than there were from Italian camps. To a very great extent and in a surprisingly paradoxical way, this reflected German efficiency and Italian inefficiency. MI9 relied on German organization to deliver, in a timely manner, coded letters and parcels containing the escape aids. Italian ineptitude, on the other hand, meant that letters and parcels were often held up for months behind a wall of bureaucracy and sluggishness. MI9, therefore, appeared to concentrate resources on aiding and supporting escapes from Germany. Time was spent in documenting the passage of the coded letter traffic. Prisoners of war with whom MI9 was in coded contact were encouraged to acknowledge receipt of the letters they received. There were a number of examples of this in John Pryor's letters. His letter dated 22/3/42, for example, contained a coded message which confirmed receipt of two coded messages from MI9

YOURS 24 DEC AND 3 JAN UNDERSTOOD . . .

The southwestern border of Germany, a potential route into Switzerland, as shown in Sheet Y (see pages 152–155).

The principal escape routes out of Germany which MI9 chose were south to neutral Switzerland and north via the Baltic ports to neutral Sweden. As Switzerland was landlocked, there had to be a further route out via France and Spain. This onward movement was through organized escape routes, among

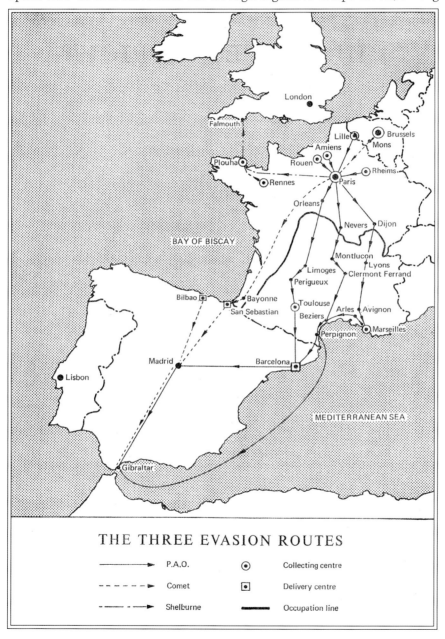

The Pat (PAO) and Comet escape lines through France to Spain are shown on this map.

the most famous of which were the Pat and Comet Lines. The Pat, or PAO, Line was named after Patrick Albert O'Leary, the cover name of a Belgian national, Albert-Marie Guérisse, masquerading as a French-speaking Canadian airman, who established the escape route from Marseilles via the Pyrenees to Madrid. The Comet Line ran through western France and, similarly, across the Pyrenees to Madrid.

Along these organized escape lines, the escapers were not on their own; they were accompanied by members of the Resistance, sometimes by paid guides and even, on occasion, by professional smugglers, who knew the terrain, so that large-scale mapping, in particular, was not required. A good example of this was coverage of the Pyrenees in [Series 43], described in Chapter 3 and listed at Appendix 5. The series covers Western Europe at a small scale (1:1,000,000) with four larger scale insets of border areas, and even those vary in scale from as small as 1:500,000, through 1:300,000 to the largest one at 1:250,000 scale. Two of the insets appear on sheet 43A and both are at 1:500,000 of sections of the Pyrenees across the French–Spanish border. MI9's mapping resources tended to be concentrated on small-scale blanket coverage with large-scale coverage of specific escape points.

The largest centres of MI6/SIS activity during World War II appear to have been concentrated in Berne, Stockholm and Madrid, the capital cities of the three principal neutral countries in Europe, and the centres through which escapers would need to travel as part of their escape routes. SIS certainly ensured from as early as the summer of 1940 that they controlled the escape routes which MI9 sought to establish through the neutral nations. It was also the case that the section within MI9 responsible for this aspect of their work was run by Jimmy Langley who was in fact a member of SIS and was never actually appointed to the staff of MI9. (This aspect of the SIS–MI9 relationship will be considered in more detail in Chapter 9.)

ESCAPE TO SWITZERLAND: THE SCHAFFHAUSEN SALIENT

The first escape route covers the Schaffhausen Salient, a rather tortuously oriented section of the German–Swiss border to the west of Lake Constance, where a peninsula-like area of Switzerland projects into neighbouring Germany. It would have been the closest stretch of the border for those seeking to escape from Germany into Switzerland. This route was important in the greater scheme of MI9 organized escape, not least in terms of the success rate which attached to the attempted escapes across this particular border: almost 20 per cent of all

OVERLEAF:
Detail of the French–Spanish border, from [Series 43] sheet A. The red line shows the actual border and the blue line shows the northern limit of the forbidden zone.

successful escapes throughout the six years of the war were into Switzerland and the majority of those were along this route.

MI9 worked hard to identify potential escape routes and provide the necessary maps to help prisoners of war to escape. One of their most intriguing maps is sheet Y, a large-scale (1:100,000) map of the Schaffhausen Salient that is very different from most of the escape and evasion maps which MI9 produced. Firstly, at 1:100,000, its scale is very much larger than most of their mapping programme. Secondly, it is clear that the localized area of coverage had been carefully selected to afford escaping prisoners of war the maximum chance of a successful escape. Thirdly, it contains a considerable amount of textual information, in military circles known as 'goings' information (terrain analysis), i.e. identifying and describing in considerable detail the significant features in the landscape which would help the escaper to navigate a successful route and also highlighting any features which would likely hamper or impede escape. The textual information on sheet Y commenced by stating:

> Escapes into Switzerland have the greatest chance of success if attempted across the frontier of the Canton Schaffhausen. The region around Lake Constance is to be avoided.

Sheet Y was based directly on native Swiss–German topographic mapping of the border area and, because of the density of detail, the colour specification and the need to print on fabric (which had initially proved a considerable technical challenge for the Waddington company), the one surviving copy of this sheet which has been identified, is quite difficult to read, despite its larger scale.

> 4.   Immediately North of the tower, there is a good chance to cross the frontier anywhere in the forests.
>
> 9.   There are also two salients of German territory, Busingen and Wiechs, completely surrounded by Swiss territory inside the Canton Schaffhausen. As soon as a fugitive is reasonably sure he is in Switzerland, he should make himself known to Swiss peasants. His reception will almost certainly be good, and the danger of wandering back into German territory would be avoided.
>
> 10.   The stretch of frontier round the Ramsen salient, and also the stretch from Erzingen westwards to the Rhine must be avoided, as barbed wire charged with electric current has been erected.

RIGHT:
Detail from Sheet Y, shown at scale. The complexity of the base mapping and the relatively poor printing registration on the only surviving copy make it difficult to read. This area shows the Canton Schaffhausen, the frontier of which MI9 advised had the 'greatest chance' of escape success. The town of Schaffhausen is just right of centre on this extract, on the north bank of the river.

LEFT:
One of the unusual features of Sheet Y was the textual 'goings' information ranged along the northern edge of the sheet, with specific advice for escaping prisoners.

OVERLEAF:
Full extent of Sheet Y, showing the Swiss–German border region (reduced to approximately 1:200,000 scale to fit on the page and with text repositioned slightly to aid clarity).

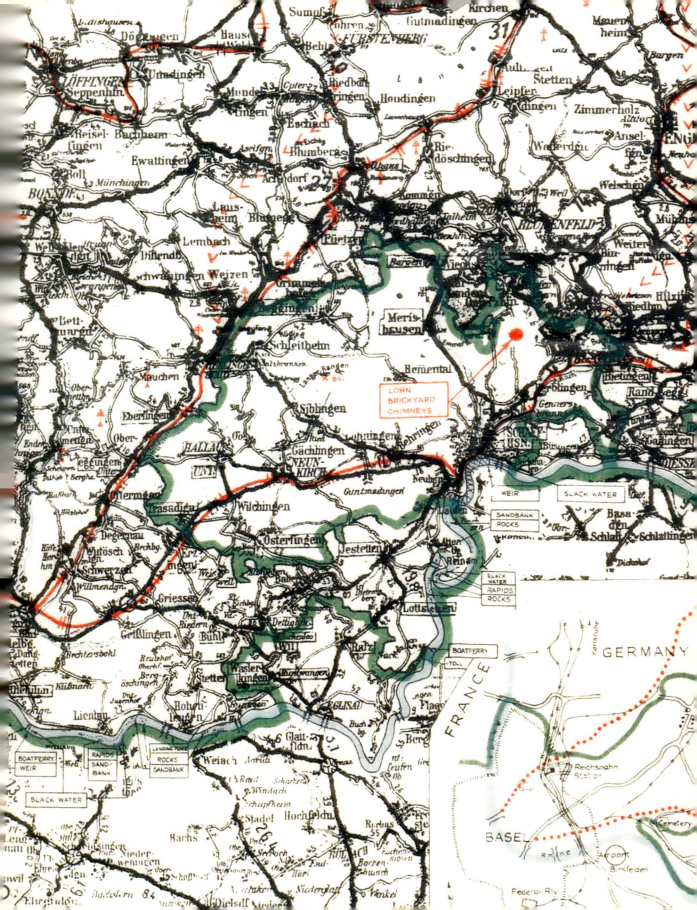

1. Escapes into Switzerland have the greatest chance of success if attempted across the frontier of the Canton Schaffhausen. The region around LAKE CONSTANCE is to be avoided.

2. A fugitive coming in a southerly direction through the Black Forest should continue until he reaches the line of overland electric cables which runs from Waldshut to Ravensburg, skirting north of the Swiss frontier. The pylons are about 70 to 75 feet high, of steel, and carry 7 electric cables.

3. Should these cables be reached anywhere on the stretch Waldshut to Thengen, it should be possible to identify an iron observation tower, standing on the top of a wooded hill. The hill itself is the highest of a chain of hills, wooded with pine and beech trees. The tower has two view platforms, both appearing above tree level and is about 100 feet high. On a clear day it can be seen from a distance of 30 kilometres in a northern or western direction. It is about 3 to 5 miles within Switzerland, and is known as the Schleitheimer Randenturm. The position is approximately

Lat: 47° 44′ N.
Long: 8° 32′ E.

It stands on the summit of a hill marked on maps as being 891 metres high. The best course would be to walk towards this until it is judged to be about 3 to 5 miles away, when the frontier will be close at hand.

The easiest point to cross the frontier lies North West of the tower, where the river Wutach, which forms the frontier, at this stretch, runs beside the railway line between Weizen station and Stuhlingen. The river runs through forest on either side, but there is a narrow belt of open fields immediately on the river banks. It would be advisable to lurk in the cover of the forest until after dark, and then cross the fields and river under cover of darkness. The river itself is narrow and not more than 3 feet deep when at its fullest. Smugglers and local inhabitants crossing the frontier secretly favour this stretch of the frontier.

4. Immediately North of the tower, there is a good chance to cross the frontier anywhere in the forests.

5. Should the electric cable line, referred to in Para 2, be reached anywhere between Thengen and Ravensburg, it should be possible to sight two volcanic hills, the Hohenstoffel, (846 M. high) position

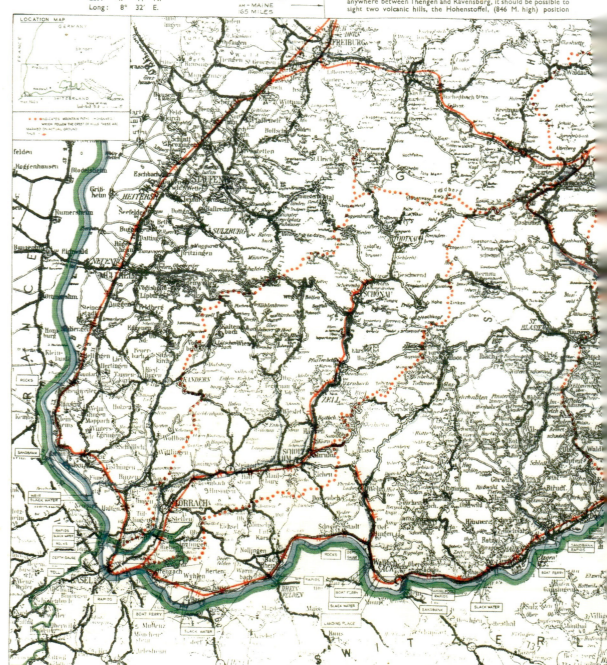

SECRET

(approx.):
    Lat: 47° 47' 30" N.
    Long: 8° 45' E.

the Hohentwil, (688 m.high), position (approx.):
    Lat: 47° 46' N.
    Long: 8° 49' E.

These two hills rise from the surrounding plain, and form landmarks visible for a distance of 60 kilometres in clear weather from a north eastern or north western direction. They are however in Germany, and they themselves must be avoided as German O.P.s are stationed there.

Having found these hills, the next point to identify would be the two parallel chimneys of the brickyards at Lohn, lying west. They are inside Swiss territory and would be the point to make for.

Should the two volcanic hills be sighted by the fugitive west of his position, he would of course have to pass the hills before he would be able to sight the factory chimneys. In this case his best chance of avoiding detection would be to pass westwards through the gap between the two hills, and then try to identify and aim for the factory chimneys.

8. On no account should the railway line Singen-Schaffhausen to the South be crossed, as the course of the frontier then becomes complicated, and it would be possible to cross into Switzerland and then immediately back into Germany through ignorance of the frontier.

9. There are also two salients of German territory, Busingen and Wiechs, completely surrounded by Swiss territory inside the Canton Schaffhausen. As soon as a fugitive is reasonably sure he is in Switzerland, he should make himself known to Swiss peasants. His reception will almost certainly be good, and the danger of wandering back into German territory would be avoided.

10. The stretch of frontier round the Ramsen salient, and also the stretch from Erzingen westwards to the Rhine must be avoided, as barbed wire charged with electric current has been erected.

11. Polish prisoners are now working on many farms and roads in South Western Germany. They wear a yellow arm band marked "POLN. KRIEGSGEFANGENER".

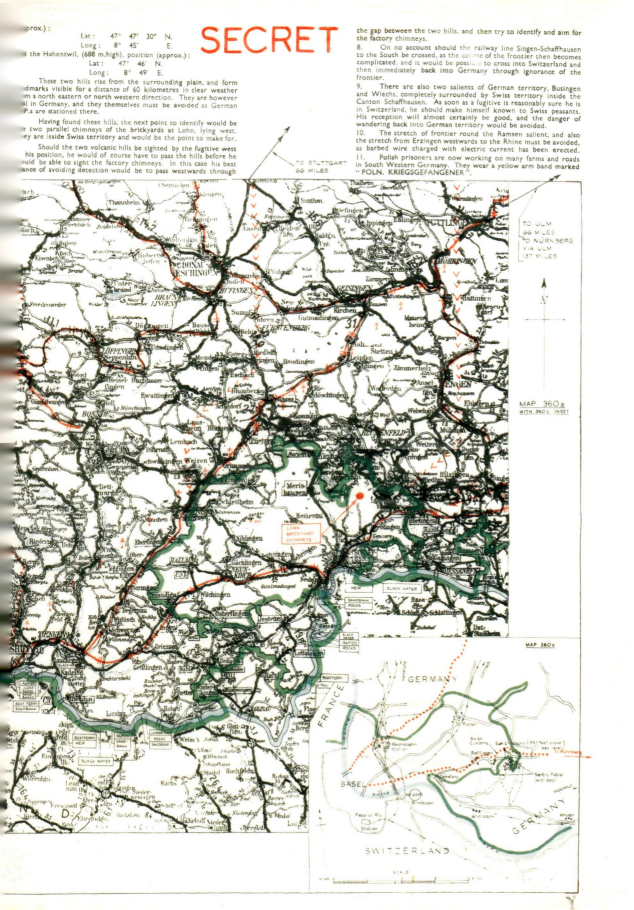

MI9 also produced cover of the same area at the same scale in two sheets, Schaffhausen Salient (West) and Schaffhausen Salient (East), marked respectively A1 and A2. Sheet A1 was produced as an escape and evasion map (see Appendix 1) but no copy of sheet A2 Schaffhausen Salient (East) escape and evasion map has ever been discovered. However, it is known to have been identical in specification to sheet A1 Schaffhausen Salient (West). A map of A2 Schaffhausen Salient (East), printed on paper, did appear in the *Bulletin*, and is reproduced here (see Appendix 9 for the *Bulletin*'s version of sheets A1 and A2). The significant difference between these two maps and sheet Y is that they were redrawn to a simpler specification to show only a selection of the topographic detail on sheet Y. Contours have been removed and elevation data were shown only in outline by hachures. The textual information has also been removed, but the man-made landmarks described have been included as annotations, for example, 'the brickyards at Lohn' on the western edge of sheet A2. The result is a considerable enhancement in the clarity and readability of the detail when compared to sheet Y.

According to the War Diary, by 8 May 1940, MI9 was despatching this 'special map of the Northern Swiss frontier near Schaffhausen with a memorandum on the prevailing conditions in the area obtained from MI sources'. This was clearly sheet Y. It is fascinating to try to discover just how MI9 was able to produce such a detailed map which contained so much local intelligence information, and key to the story is the role played by Johnny Evans.

His role as a lecturer, user of coded correspondence as a means of escape, and source of ideas for escape aids, has already been discussed, particularly in Chapter 1, but it was really in the production of sheet Y that his contribution appears to have been even more significant. Evans had been a Major in the newly formed Royal Flying Corps in World War I. He had been involved in the Somme offensive in July 1916, overflying the German front line in an attempt to take out their gun batteries. His aircraft engine failed and, although he survived the crash landing, he was immediately captured. Because his brother had been captured at Ypres, and the family had received no news of him for over a year, together with the statistical likelihood of capture as a member of the fledgling Royal Flying Corps, Evans had had the foresight, prior to his deployment, to set up a coded contact system with his mother. His mother was able to maintain contact with him once he had been captured and she succeeded in providing maps and compasses, secreted inside food parcels, to aid his ultimately successful escape.

Evans, and his colleague, Lieutenant S. E. Buckley, originally with the Northamptonshire Regiment and later with the Royal Flying Corps, escaped

TOP:
Sheet A1 Schaffhausen Salient (West), printed on silk.

BOTTOM LEFT:
Sheet A2 Schaffhausen Salient (East) printed on paper and taken from the Bulletin.

BOTTOM RIGHT:
Detail from Sheet A2 showing the Swiss–German border at a crossing point that was used by several escapers including Neave and Luteyn during their escape from Colditz (see page 160).

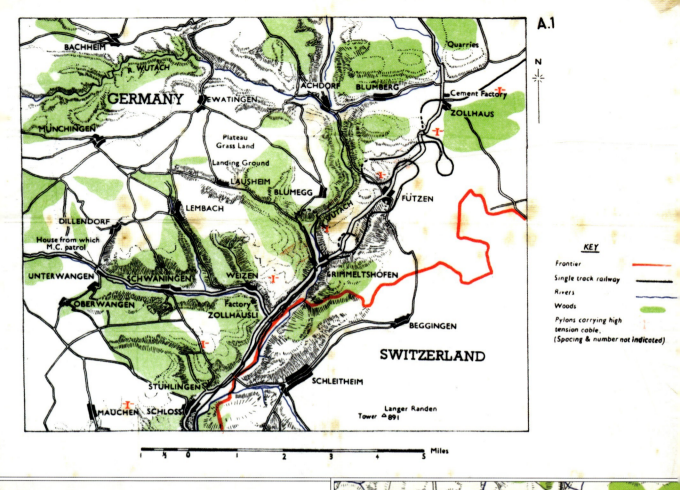
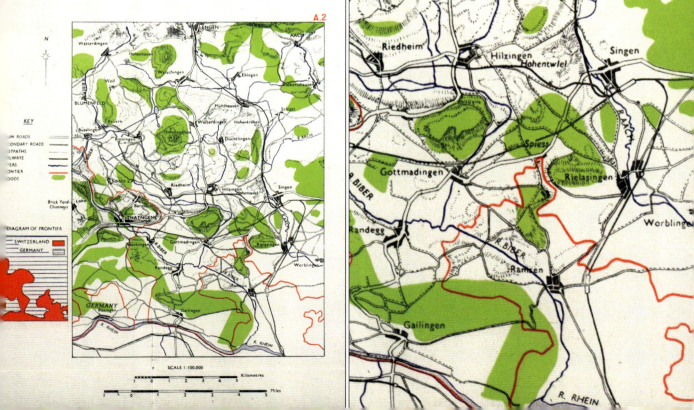

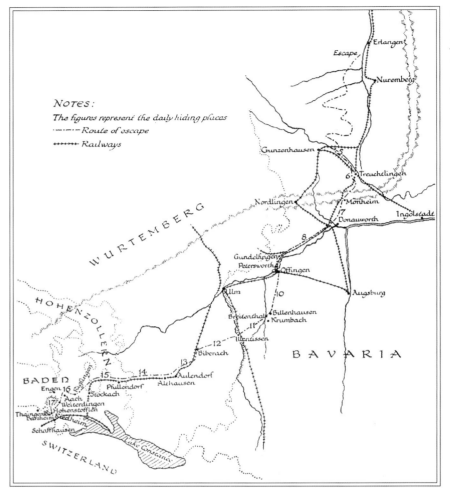

A sketch map of the escape route of Johnny Evans from near Nuremburg to Switzerland in 1917, from his account, The Escaping Club, *published in 1921.*

from a train whilst being moved from one camp to another. They walked over 200 kilometres (130 miles), south from the point where they jumped from the train close to Nuremburg, all the way to the Swiss border. They had decided not to attempt to board a train but rather rely on walking to the border and staying off the beaten track. It took them eighteen days to reach their objective, the Schaffhausen Salient, and cross safely to freedom in Switzerland. Their biggest challenge was to find enough food to sustain them and they resorted on many days to digging up potatoes from the fields they passed as the meagre rations they carried had run out early in their escape. They did, however, have the 'excellent maps' which Evans' mother had sent him and 'accurate and detailed knowledge of the whole route . . . to the frontier' acquired from fellow prisoners who had escaped, been re-captured and had briefed fellow escapers on the detail of the route to freedom. However, Evans clearly recalled the problems they had faced on this particular stretch of the border and, specifically, in knowing precisely when they had reached safety. On such a winding stretch of the border, it was easy to be confused and cross back into Germany by accident. The detailed textual information which appeared on the northern edge of sheet Y clearly reflected his own, first-hand experience, as follows:

> On no account should the railway line Singen–Schaffhausen to the South be crossed, as the course of the frontier then becomes complicated, and it could be possible to cross into Switzerland and then immediately back into Germany through ignorance of the frontier.

Evans became convinced that his World War I experience might prove of value during a second war which many began to anticipate in the 1930s.

Evans, however, went further than simply recalling his World War I experience and offering the benefit of it to MI9. As Hutton described his actions, he decided to spend a holiday cycling around the Schaffhausen district in 1937. His objective was to photograph both sides of the section of the German–Swiss frontier across which he had made his 'memorable march to freedom'. The fact that sheet Y carried detailed textual directions on the nature of the frontier alignment doubtless reflected his own experience. Evans subsequently joined MI9 in 1940. While no record has been found which clearly identifies sheet Y as the product of Evans' foresight, there is much circumstantial evidence to support such a hypothesis. It is a matter of record that the map covers the same stretch of the German–Swiss frontier across which he had escaped to freedom in World War I; he produced much of the intelligence described in minute detail on the map; he was certainly a committed member of the MI9 staff from the beginning; the map of the Schaffhausen Salient was one to which MI9 attached great significance and Evans was, from the beginning, regarded as one of their best lecturers at the Training School. Hutton reinforced the perception of Evans' contribution by emphasizing the extent to which Evans' photographs of the Schaffhausen Salient proved invaluable to MI9's work. These photographs are likely to be the ones reproduced in the MI9 *Bulletin* showing detailed views of the topographic and landmark features which would be of most navigational significance to those attempting to escape across this particular stretch of the frontier, including views of river banks and areas patrolled by German border sentries.

There were many escapes through the Schaffhausen Salient which were documented in the records, almost all of which were initially highly classified. Pryor had asked for maps of the Swiss border in one of his coded letters and he also mentioned in his memoirs the escape attempt by Lieutenant R. F. Jackson on 2 February 1944. Jackson was fluent in French and was furnished with papers indicating he was a French national. Having succeeded in getting clear of Marlag and Milag Nord camp, he travelled south by train to Schaffhausen, a considerable distance from the camp in northern Germany, and then made his way on foot to the border. Unfortunately, in the dark, he stumbled on a trip wire and was caught by guard dogs before he was able to cross into Switzerland. It is notable that he chose to travel south to the Swiss border, which had been the area for which Pryor had originally requested maps in 1942. Whether or not Jackson had access to maps is not mentioned by Pryor. The importance of the route and of the maps produced of the area is shown in the story of the escape of Airey Neave from Colditz.

RIGHT:
The prisoners' courtyard at Colditz.

AIREY NEAVE'S ESCAPE FROM COLDITZ

Airey Middleton Sheffield Neave was born in London on 23 January 1916. Educated at Eton, he went on to read jurisprudence at Merton College, Oxford, where he graduated in 1938. At the outbreak of war, Neave (already in the Territorial Army) became a lieutenant in the Royal Artillery. He was sent to France early in 1940 as a Troop Commander in the 1st Searchlight Regiment of the Royal Artillery. On 26 May, whilst trying to defend a forward position to the south of Calais, he was wounded and subsequently captured, as he lay on a stretcher. He was held in various oflags and, after a number of unsuccessful escape attempts, he was eventually imprisoned in Oflag IVC, the castle in Saxony more commonly known as Colditz. The fortress, which stood on the site of an earlier castle, had been largely constructed in the sixteenth century by the Elector of Saxony, and extended further by Augustus the Strong of Saxony in the eighteenth century. It was to Colditz that those who repeatedly attempted to escape were often sent as its construction and location ensured that breaking out of such a fortress was a considerable, if not impossible, challenge. Neave was in fact the first British officer to escape successfully from Colditz, and not Pat Reid who was depicted as such in the film *The Colditz Story*. Neave told his own story in *They Have Their Exits,* published in 1953. He insisted that he had written the book without access to official sources, but simply from his own personal recollections. That may well have been the case but it is certain that

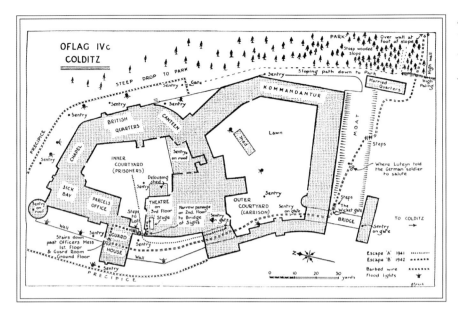

A plan of Colditz, showing the route of Airey Neave's successful escape (ESCAPE 'B' 1942), from his account, They Have Their Exits.

he would still have needed official sanction at that time to publish such a book, not least since, as a recently elected Member of Parliament, he was very much in the public eye.

In many ways Neave personified the approach to escape which MI9 had sought to inculcate. He mentioned the philosophy of escape a number of times in the published account of his escape in a very telling way, to the extent that it was later quoted by Foot and Langley:

> The real escaper is more than a man equipped with compass, maps, papers, disguise and a plan. He has an inner confidence, a serenity of spirit which makes him a Pilgrim.

Neave had attempted to escape on two previous occasions from Colditz, on 20 August 1941 and on 23 November 1941. The first attempt saw him try to walk out dressed as a German guard and the second involved crawling through the attics and trying to drop down seventy feet on a knotted sheet over the wall: both attempts failed before he got clear of the camp. He eventually succeeded on 5 January 1942. Dressed as a German Oberleutnant, he was accompanied by his Dutch colleague, Lieutenant Toni Luteyn of the Netherlands East Indies Army. The two nationalities had apparently 'agreed to pool their resources for escaping' according to Foot and Langley. Most of the Dutch officers spoke fluent German and were, therefore, ideal travelling companions. The British were able to contribute the aids to escape which had been provided by MI9.

In plotting the entire route of their escape, as described by Neave in his book, on a small-scale map of Germany and northern Switzerland and also using his subsequently discovered escape report, one aspect of their escape became abundantly clear. While he and Luteyn travelled south towards the Swiss frontier, they did not make for the nearest point of the frontier, i.e. they did not travel due south to the frontier but veered south west at one point in order to reach the Schaffhausen Salient.

From Colditz, they walked over nine kilometres (six miles) to the small town of Leisnig where they caught a train to Leipzig, a distance of some forty-eight kilometres (thirty miles) further on. From there they travelled south 290 kilometres (180 miles) by train to Regensburg where they changed trains for Ulm. They travelled a further 177 kilometres (110 miles) to Ulm where, on

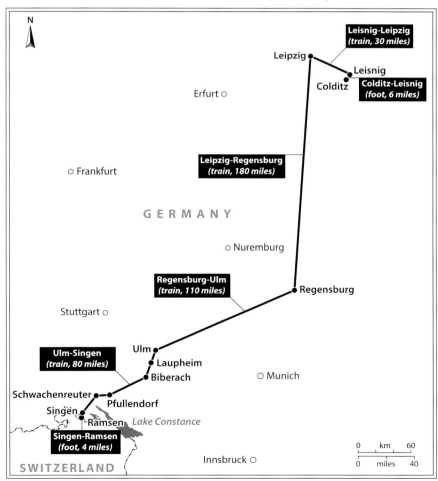

Neave's successful escape route from Colditz to Switzerland.

THE SCHAFFHAUSEN SALIENT AND AIREY NEAVE'S ESCAPE   163

A group of escapers in a flat on Quai Rive Neuve, Marseilles, in 1942: Francis Blanchain, Mario Prassinos, Hugh Woollatt, Airey Neave and Louis Nouveau. This was clearly taken after Neave and Woollatt (who had escaped from Oflag VB (Biberach)) had arrived from Switzerland and before they departed for Spain and the journey home.

arrival, they again changed trains, trying to get to Singen on the Swiss border. Singen is located some 130 kilometres (80 miles) southwest of Ulm, close to the Swiss border and the Schaffhausen Salient, and to the west of Lake Constance. At Ulm they had almost been recaptured since their attempt to purchase tickets for a destination on the Swiss border had apparently aroused considerable suspicion. Having avoided recapture, they then decided to continue to travel by train to Singen, but to avoid the main line, travelling rather through Laupheim, Biberach, Pfullendorf and Schwachenreuter. From Biberach it would arguably have made more sense for them to travel directly south towards the eastern end of Lake Constance, being less than sixty-five kilometres (forty miles) from the Swiss border, instead of which they chose a much longer route They eventually crossed the frontier south of Singen, reaching Ramsen which was the first settlement inside Switzerland.

In considering their route to freedom, which Neave eventually described in more detail in his escape report, the virtually inevitable conclusion is that Schaffhausen was always their objective. Why would they choose that route if they had not already been briefed on the most likely point to cross the border successfully and consulted a map which showed detailed coverage of that stretch of the border?

Such a hypothesis is also supported by independent evidence. While this may never be proved conclusively, it seems highly likely that they had access to sheet Y and the intelligence contained on it, or to sheet A2, since both would have provided them with the detail of the area they needed and demonstrably possessed. It now appears likely that Neave was trying to hide the fact that such a map had actually been sent covertly into Colditz by MI9, since he mentioned it neither in his book nor in the escape report written on arrival safely back in London. Neave's apparent reticence with regard to the role played by possession of a covertly acquired map reinforces Professor Foot's insistence, in a conversation in 2012, that 'maps were never discussed' as that is what 'we had been briefed on'. It is also significant that Neave's account was published in 1953, less than ten years after the end of the war, at which point the protocols and restrictions regarding public disclosure of MI9's methodology would still have been paramount. The Korean War did not end until the summer of 1953 and it is possible that military maps were still being produced on fabric at that time, although probably intended more for operational use rather than for escape and evasion purposes. A 'D' Notice (number 42) had been issued on 19 January 1946 which forbade the public disclosure of escape and evasion methods, including any assistance given to prisoners of war. The Notice also embraced the non-disclosure of information relating to the use of secret methods of communication with the prisoner of war camps and the identities of nationals of European countries who had assisted in the escape programme. The system of 'D' notices was introduced in 1912 and still continues. Defence Advisory Notices advise organizations in the UK on which information should not be published or broadcast, since to do so could prove prejudicial to the national defence interest.

There is, however, quite specific evidence to confirm that MI9 had, in fact, sent advice into Colditz on some declared escape requirements 'concerning routes and destinations', and that they had not only made plans to send parcels containing escape materials to selected officers in a number of specified camps, of which Oflag IVC (Colditz) was one, but that many such parcels had already been sent by the time of Neave's escape. By May 1941, MI9 had received confirmation that escape material had been received in four camps, of which Colditz was one of those listed. During the same month, MI9 had despatched sixty-two parcels under the guise of the Prisoners' Leisure Hours Fund. By July 1941, parcels had certainly arrived in Colditz, confirmation having been received in MI9: it was noted that they had been despatched in March and April 1941.

Additional evidence that it was indeed either sheet Y or A2 which Neave and Luteyn used is also to be found in the history of the Colditz camp. After the war had ended, and on the repatriation of the prisoners of war, the Senior British

Officer (SBO) in each camp was required by the War Office to produce a written record as an historical review of life in the camp. It is likely that these reviews were produced in order to ensure that any possible war criminals among the German guards, or treasonable behaviour by any of the prisoners of war, could be documented and pursued through proper legal process. In writing the Historical Record of Oflag IVC (Colditz), the rapporteur included, at Chapter X, a review of the successful escapes. The escape report written by Captain P. (Pat) R. Reid of the Royal Army Signals Corps and Flight Lieutenant H. N. Wardle of the RAF was included. Reid and Wardle escaped from Colditz on 16 October 1942, nine months after Neave and Luteyn. Following an identical route, they made for the same crossing point on the Swiss border and, once successfully across, they turned themselves over to the Swiss border guards in the same town, Ramsen, as Neave and Luteyn had done. Their report includes far more detail about the maps they used, including mention of a 'Swiss frontier map and half inch diameter brass compass, both W.D. issue.' In this context, W.D. almost certainly referred to the War Department (or Office). It also becomes apparent that they must have received feedback from Neave through the coded letter system since they made very specific mention of beginning reconnaissance in daylight 'to find Neave's fork' in the road leading to the frontier. Many years after the war, in 1974, Neave himself confirmed this in an interview he gave to his local newspaper. He stated unequivocally that he had sent back details of the precise Swiss frontier crossing point to Colditz in a coded letter and that 'the same route was later used successfully by Major Pat Reid'.

The SBO's history also made very clear that 'a comprehensive supply of maps covering the whole of Germany was available to intending escapers'. Those men whose escape plans had been endorsed by the Escape Committee were required to make their own copies of the maps so that they would learn more thoroughly the detail of their planned route. It was also made clear that intending escapers were shown detailed maps of the frontier but were never allowed to copy those particular maps, being required rather to memorize them, for reasons of security.

Careful reassembly of the various pieces of the story has shed important new light on Neave and Luteyn's chosen escape route. It can now be reasonably asserted that such parcels as had already been despatched to Colditz prior to January 1942 contained specifically a copy of sheet Y or A2 which aided considerably the first successful escape of a British officer from the camp, that of Neave. It is, moreover, now possible to contend that, although Neave described the route of his escape in considerable detail in his published account and said next to nothing about the maps, he nevertheless made use of them. In *They Have Their Exits*, he described how he had traced in Indian ink 'the neighbourhood of the Swiss frontier from a stolen map', offering no explanation on the apparent theft. To have had an opportunity to steal a map of the Swiss border is, at best, regarded as unlikely, not least since Colditz was located in central Germany, a considerable distance from the Swiss border. He added that the currency they were given for their journey had come from 'black market deals with the guards' and that Luteyn had been able to buy 'a map of the surrounding country in a small shop' in Ulm during their journey to the border. That again appears to be an unlikely assertion.

There is, however, documentary evidence which indicates that MI9 had despatched gramophone records to eighteen named individuals in Colditz in May 1941. The despatch list indicated that one record in each box contained 'a map of the frontier'. While there is no indication that it was actually sheet Y or A2, or indeed that the frontier was the Swiss frontier, the circumstantial evidence is strong. One of those parcels was addressed to Captain P. R. Reid (see page 96). Reid had been appointed in January 1941 by the Colditz SBO to be in charge of escaping. By October of that year, an Escape Committee had been formed and Neave had been appointed as Reid's deputy and had been placed in charge of the Committee's Maps Section. It was the Map Officer's responsibility to hide the maps being held in the camp. It was Reid who closed the door through which Neave and Luteyn escaped from Colditz, and who was subsequently able to escape successfully via the same route.

Neave's personal papers contain a sketch map showing the route he and Luteyn took once they had arrived in close proximity to the Swiss border. It was surprisingly, however, not drawn by Neave, but was rather sent to him by a Cuban architect, Roberto Pesant, who wrote in the accompanying letter, dated 23 September 1958, that it was his attempt to reconstruct Neave and Luteyn's crossing of the Swiss frontier from the textual description contained in Neave's book, published in 1953. He asked that Neave comment on the accuracy of his tracing of the exact escape route. Neave replied to Pesant on 9 October that the sketch map was 'extremely good', correcting a couple of aspects and commenting 'we actually went through the town of Singen travelling westward before we turned south to cross the frontier as your map describes'. Both sheets Y and A2 show the frontier crossing point on the road between Singen and Ramsen, located in the extreme southeast corner of both sheets, but depicted with rather more clarity on A2 than on sheet Y.

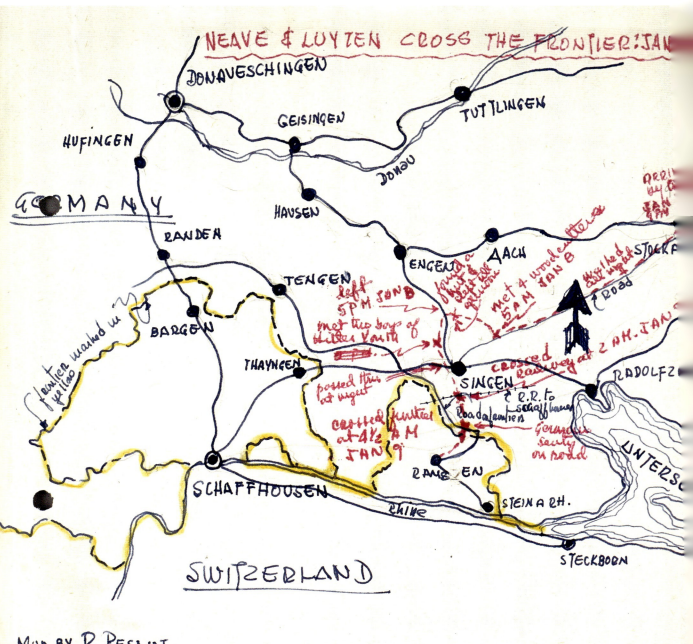

LEFT:
The route taken across the frontier by Neave and Luteyn, as reconstructed by Roberto Pesant in 1958, based on Neave's book, They Have Their Exits.

For MI9, any escape from Colditz was a priority. These were, after all, the men who had repeatedly tried to put the MI9 philosophy of escape-mindedness into action. They had failed to make it back to the UK to date, but they had certainly proved themselves worthy of every aspect of the support mechanism which MI9 could muster. To this end, MI9 had developed a special coded message system, Code V, and proceeded to reserve its use for contact with the Colditz prisoners of war. They managed to send a detailed explanation of the official system through a series of coded messages using the existing private code which one of the prisoners of war, Captain R. F. T. Barry, had had the foresight to set up with his wife prior to deployment. It is also clear that Neave was on MI9's Special Watch List since the War Diary entry for January 1942, the month in which Neave escaped, noted that during the month 'one British Officer succeeded in escaping from Germany to Switzerland. He was on the Special Watch List and had received escape material'. Whilst he was not identified in the War Diary entry, it was rare for the War Diary reports to comment on individual escapes; they tended rather to document monthly escape statistics. Since Neave was the first British officer to escape successfully from Colditz, clearly using information with which he had been provided by MI9, it was likely regarded by MI9 as worthy of special mention.

Like all successful escapers, Neave was interviewed on his return to London. The report of his debriefing is of particular interest in itself. The report is not to be found in the general collection of such reports where the vast majority are to be found but rather in the Historical Record of MI9, almost buried from sight. A copy was also subsequently found in the Colditz camp history. The report was also notable for its brevity. Most of the reports carried detailed descriptions of the escape, offering commentary on likely routes, appropriate behaviour, and places to avoid. There is a far more detailed description of the escape in Neave's published account than there is in the debriefing report. Certainly he made no mention at all in his escape report of any use of maps. It was MI9's practice to ensure that any information gleaned from returning escapers was relayed back to the camps to help inform and update the planning of the escapes. It was for that reason that a coded message was relayed on 11 November 1944 indicating:

> STRANGERS NEAR SCHAFFHAUSEN SALIENT ARE LIABLE TO ARREST FOR QUESTIONING EVEN IF PAPERS ARE IN ORDER

While it is known that the message definitely went to Marlag and Milag Nord camp, such a message would certainly have been relayed to all the camps with which MI9 was in contact. Some detail of the Swiss frontier was also acquired from the Dutch since two of their number had almost succeeded in reaching Switzerland, only to be foiled at the last moment, recaptured and returned to Colditz.

There is clearly sufficient evidence to support the hypothesis that Neave indeed had access to sheet Y or A2, although he personally never declared that fact in his description of the escape, either in the report he wrote on arrival back in the UK in 1942 or, indeed, in the book he subsequently wrote about his escape. Neave kept his secrets, as he had been briefed to do, and it is understandable that no inkling of what really happened in terms of the maps was ever revealed in his book, published so soon after the war. He certainly made clear in the interview he gave in 1974 that there was a great deal he had never revealed. Sadly, he was murdered by the Irish National Liberation Army in 1979, long before the notions that secrecy about possession of escape and evasion maps would have seemed redundant. Even so, it is worth recalling that Foot remained reticent in discussing the maps until the end of his life.

Airey Neave took most of his knowledge on MI9's escape maps to the grave when he was murdered by the Irish National Liberation Army with a car bomb that exploded outside the Houses of Parliament in 1979.

This all certainly appeared to mirror the pre-deployment briefings which Foot had described and the order that 'maps were never discussed', and the extent to which it reflected his and Langley's expressed view that to discuss such methods could prove prejudicial to their future deployment. Looking at the statistics produced by MI9 at the end of the war, it is clear that the escape route out of Germany and into Switzerland was not only one of their targeted routes but was also one of the most successful: the total number of escapes through Switzerland was 5,143, almost 20 per cent of all successful escapes from enemy occupied territory by the end of the war. Foot and Langley reasonably drew attention to the fact that wartime statistics 'cannot be claimed as perfect', and there are certainly variations in the figures which appear in various sources at the time. The figures do not, however, vary significantly enough for there to be serious concern over their veracity. It is also notable that the figures did emanate from MI9 which, it is known, was keenly and carefully monitoring the numbers of escapes on a monthly basis.

Unravelling the story of Neave's escape has served to bring into focus the true contribution of the mapping programme to successful escapes and emphasizes the extent to which MI9's considerable efforts to communicate with the camps and smuggle in the maps paid very real dividends. Escaping from Colditz was, in itself, a significant challenge; travelling successfully through Germany around 650 kilometres (over 400 miles) to the Swiss border and crossing successfully to freedom was arguably only possible with detailed navigational information which the maps and the supporting intelligence provided.

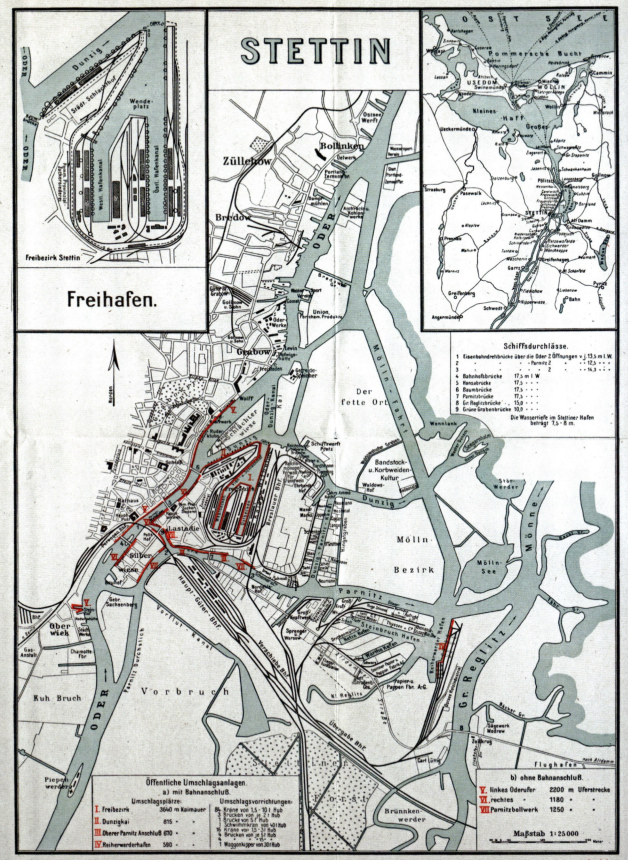

7

ESCAPING THROUGH THE BALTIC PORTS

'I saw the S.S .BJORN, STOCKHOLM moored in the Swedish dock and being loaded with coal . . . as regards local geography, the flimsy of which I had a copy, proved invaluable . . .'
(Account of the escape of Flight Lieutenant Oliver Philpot)

The route via the Baltic ports was another of MI9's recommended escape routes and was to prove even more successful than the Schaffhausen route. The *Bulletin* made clear that it was the most effective route to try to escape from occupied Europe and reach Sweden, a neutral country. Danzig, Gdynia, Stettin and Lübeck were regarded as the key Baltic ports targeted by MI9 as potential escape routes and the aim was always that the escapers should be looking to board a ship of the neutral nation, Sweden. *In extremis*, some escapers resorted to Finnish ships in the hope that they planned to dock in a Swedish port on their return trip to Finland. Large-scale maps of all four ports were produced for incorporation in the *Bulletin* for training and briefing purposes.

There is also separate evidence to support the contention that MI9 produced escape and evasion versions of the port plans. A large-scale map of Danzig harbour (A4) at approximately 1:16,000 scale was produced in relatively small numbers in October 1942, some 300 copies in total, half of which were printed on silk and half on paper (see page 50). Analysis of the detail has shown that it was based directly on British Admiralty (BA) chart 2377. The BA chart actually comprised two large-scale port plans of Danzig and Gdynia harbours. The Danzig plan indicated its primary source as a 1933 German Government Chart. Sheet A4 was a small section of just the BA Danzig port plan, centred on the railway terminal and timber wharves at Wechselmünde, but oriented differently, being significantly east of a true north bearing, and it was void of all soundings information. The basic topographic detail was retained and enhanced with what can only be described as field intelligence annotations.

The Danzig map was produced in various forms by MI9. At least three have been identified: there may have been more. The different versions, varying in geographical extent, content, scale and sheet numbering identification, are listed at Appendices 1 and 9. Close scrutiny of the detail revealed the maps' true purpose. Almost in the centre of sheet A4 were two annotations: 'Swedish ships load coal here' and 'Swedish ships unload ore here', while sheet A3 has two shortened annotations at the same point saying 'Swedish ships'. If an escaper

LEFT:
A 1:25,000 scale map of Stettin, produced for the Bulletin. *Stettin was one of the four Baltic ports recommended by MI9 as exit points for escape to Sweden.*

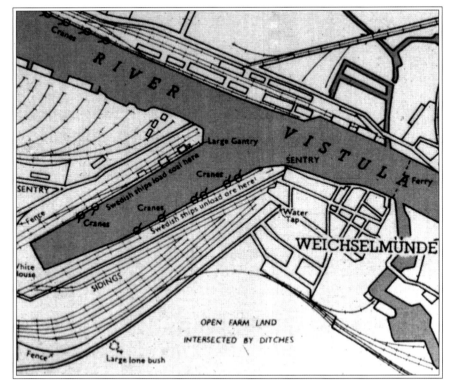

Detail from the escape and evasion map of Danzig, (sheet A4) identifying where Swedish boats loaded coal and unloaded ore. For an image of the complete map see page 50.

could board the ship of a neutral nation, his escape had virtually succeeded. Should the escaper get as far as the port, there was sufficient information on the map not only to locate ships of a neutral nation but also to show the most accessible route to the wharves where those ships would have docked. Additional annotations showed 'open farm land intersected by ditches', 'impassable marshy ground', 'large lone bush', the location of sentries and searchlights, with even the extent of 'the possible arc of light' being drawn in the area of the railway sidings. Such detailed intelligence could only have been acquired from returning escapers who had successfully navigated that escape route. All who successfully returned to the UK were questioned closely on their experiences. The resulting reports described the initial escape from the camp and the route to freedom, and offered advice based on the successful escaper's experience. All relevant intelligence was fed back directly to the camps, through the coded letter system, was included in the training lectures and the *Bulletin* and almost certainly reflected in relevant escape and evasion mapping whenever possible.

The small numbers produced of the Danzig port plan appeared to indicate that MI9's intention was never to issue them directly to potential escapers

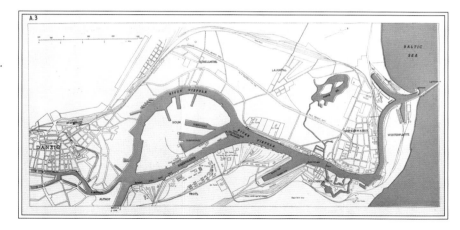

At least three maps of the port of Danzig were produced by MI9. This is sheet A3, from the Bulletin.

during escape training or pre-deployment briefing but rather to use them as briefing aids and also to send them directly to the camps for targeted escapes. Certainly, of the four port plans which appeared in the *Bulletin*, surviving escape and evasion copies of only one of the port plans, Danzig, have been discovered to date. No copies of Lübeck, Gdynia or Stettin produced on fabric have yet been found and none of the production records identified to date has any indication of their production on fabric or paper media as escape and evasion maps. However, entries in the War Diary do indicate that escape and evasion versions of some of the port plans were produced. For example, in December 1942, the War Diary entry indicated that a sketch map of Stettin Docks had been passed for reproduction prior to being sent to the German camps. Similarly, in February 1944, the War Diary entry reported that a plan of Gdynia, described as being 'maps for prisoners of war', had been passed for reproduction. It is, therefore, clear that the port plans were being produced as escape and evasion maps to be sent directly to the camps in Germany to support planned escapes.

It should be recalled that those being briefed prior to operational deployment or during training at the Training School were not allowed to make copies of any maps which appeared in the *Bulletin*. Those intelligence officers who were briefed on these potential escape routes were allowed only to use the maps for briefing purposes and those being briefed were expected to memorize, and not copy, the map detail. The evidence of the War Diary entries does, however, appear to indicate that the large-scale plans of the Baltic ports were being produced and sent to the camps, although there is no indication of the numbers being printed. It is beyond question that these routes out of captivity to freedom were an integral part of MI9's plans. They were well used by escapers and many succeeded in gaining freedom through the Baltic ports.

OVERLEAF:
Northern Germany and the Baltic coast, from [Series 43] sheet E. Three of the vitally important Baltic ports, Stettin, Danzig and Gdynia, are shown on this map.

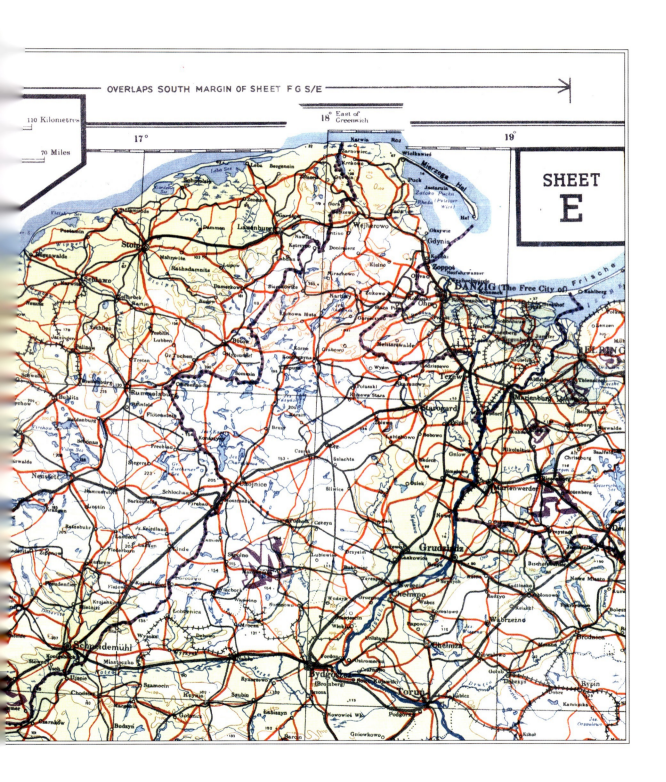

ESCAPING THROUGH THE BALTIC PORTS

Sweden proved to be the most successful route home for many escapers and evaders: 30 per cent of all returning personnel came home this way and the large proportion had travelled there via the Baltic ports. The advisory notes on escape routes in the *Bulletin* were regularly updated with fresh intelligence and tips from returning escapers. In February 1944, the *Bulletin* was revised to indicate that contact with foreign dockers could be made in certain Stettin cafés, and there were many helpful suggestions on the best ways to stow away successfully on board Swedish ships. Coded letters were also used as a means of conveying escape intelligence and advice to the camps. For example, in September 1944, a coded message was sent by MI9 to Lieutenant Commander M. J. A. O'Sullivan RN in the Marlag and Milag Nord camp indicating that four escapers had reached Sweden via Stettin. It described that:

> TWO ENTERED GOTZLOW QUAY WHERE WIRE MEETS SE END OF BASIN. TWO GOT SHIP REIHER WERDER HAFEN. HELP RECEIVED FROM FRENCH CAMP JUST E OF ZABELSDORF STATION

Barely a month later, however, a coded message was sent to the camps indicating:

> REF ESCAPES NO MORE SWEDISH SHIPS FROM HUN PORTS

a clear indication that, for whatever reason, Swedish ships were no longer calling at German ports. This was likely to have resulted from Swedish anticipation of allied victory over Germany, as they advanced northwards and eastwards from the Normandy landings. Effectively, this meant that the escape route via the Baltic ports was no longer feasible. Although the order that prisoners of war need no longer consider it their duty to escape was not issued until 13 January 1945, the camps certainly knew of the Allied landings on mainland Europe well before that date (not least through regular wireless contact).

ATTEMPTED ESCAPES VIA THE BALTIC PORTS

Three escapes which attempted to use the route via the Baltic ports will now be examined, one unsuccessful and two successful.

Lieutenant John Pryor RN

The first is that of Lieutenant John Pryor RN, who has already been mentioned in Chapter 5, since it was his coded letters which were used as an example to demonstrate their role in MI9's work. Pryor's experience also highlights other important aspects of MI9's escape programme and escape philosophy, and the significance of the Baltic ports as an escape route.

Pryor was educated at Oundle School, Northamptonshire, from 1933 to 1937. He took the military candidate examinations in 1937 and joined the Royal Navy in September that year. As a cadet he trained on HMS *Erebos* and then served as a midshipman on HMS *Vindictive*, HMS *Hood* and HMS *Warspite*. In 1940 he was ordered to Poole to assist in the small ships' evacuation from Dunkirk. On Thursday 13 June, he was on board an unarmed Dutch coaster, HMS *Hebe*, as second lieutenant, and played his part in rescuing men from the Normandy beaches, transferring them to a troop-ship waiting off-shore. On returning to shore to collect a further group, the coaster grounded on an unmarked shoal and came under heavy fire. Everyone on board was forced to disembark and were then captured by the Germans. As a young officer, and in line with German practice to separate officers and men, Pryor was separated from the crew of the *Hebe* immediately and grouped with some captured British Army officers. Two of these immediately approached him, asking if he would join them in an escape attempt. MI9 always briefed that the best time for prisoners of war to escape was before they finished up behind barbed wire. Sadly the escape attempt did not include Pryor as he was quickly separated from the Army officers and placed in another lorry for transport away from the coast. The group he was in was moved in stages by road, north to Belgium, and by late June they were travelling on Dutch and German waterways deep into the heart of Germany. By 6 July he was in Laufen, southeast of Munich having travelled for some days on board a train crowded with British prisoners of war. In Laufen, close to Austria, he was taken by lorry to Oflag VIIC/H, a camp converted from the Archbishop of Salzburg's summer residence on the west bank of the River Salz, where some 500 captured British officers were being held.

It was another month before his parents heard that he had been captured and he received his first letter from them on 24 September, dated 12 August. He 'celebrated' his birthday on 12 November which he recalled, unsurprisingly, as a 'bad day'. His memoirs do, however, highlight the extent to which the prisoners of war showed that indomitable British sense of humour and tried to poke fun at their German captors, a practice which they often described as 'goon-baiting'. As the Germans attempted to make them drill to German words of command, they would respond with everyone doing different things, ensuring that the entire parade became a shambles. On 18 November all the seafaring officers, whether from the Royal or Merchant Navies, were forced to march twenty kilometres

Prisoners of war at Marlag and Milag Nord camp, July 1943. John Pryor is fourth from the left in the third row from the back.

(twelve miles) to Oflag VIIC/Z at Titmoning. It was here, he recalled, that they started to receive Red Cross parcels.

The story of how Pryor joined the team of MI9 coded correspondents while in Oflag VIIIC/Z has already been told in Chapter 5. On 20 January 1941, Pryor, Elder and other naval officers were transported by rail and lorry to the Sandbostel camp in northern Germany, some forty kilometres (twenty-four miles) north east of Bremen. This proved to be a large camp comprising several barbed wire compounds, rows of military huts and watch towers. The flat, sandy topography must have looked particularly bleak in the middle of winter, but it was clear that they felt more comfortable surrounded by naval colleagues in the Marlag and Milag Nord camp huts.

Pryor's awareness of the extent to which 'escaping was the duty of a PoW', however hopeless a task it might appear, has already been highlighted. Both in this camp, however, and the new Marlag and Milag Nord camp later constructed by the Germans southwest of Sandbostel, it is clear that there was always much activity focused on escaping, specifically on digging tunnels. Additionally, a radio had been acquired in exchange for 2,000 cigarettes from a Belgian worker. The prisoners of war were often better provided with cigarettes and chocolates from Red Cross parcels, bolstered also by supplies sent in by MI9, than either the local German population or conscripted workers, and they often used these as a form of currency. The new radio replaced the crystal set, which had been ingeniously constructed by using the wiring of the hut as an aerial and which had been accidentally destroyed by the German guards during a search. This

Everyday life at Marlag and Milag Nord camp. A line of men stand in a queue outside a low wooden hut to see a dentist at the camp hospital, drawn by John Worsley, April 1944.

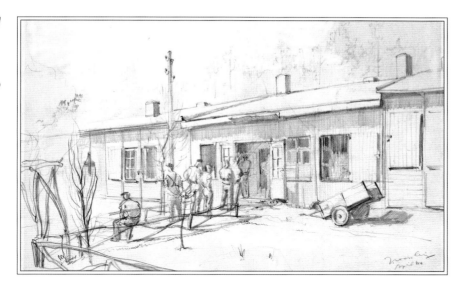

allowed them to listen to the BBC news which was subsequently cascaded orally through the other huts every night.

Pryor had been sending coded messages to his parents for some time but it was on 7 May 1942 that he sent the one requesting specific help from MI9 in support of a planned escape. The deciphered message, requesting, amongst other things, maps of the Swiss border has been described in detail in Chapter 5, as has the parcel he received that contained a chess set with escape aids hidden within it. Together with Lieutenant John Wells, John Pryor's plan to escape, which would have been sanctioned by the Escape Committee, was for them to pretend to be foreign workers dressed in suitable working attire and carrying forged passes. They planned to travel by night and hide by day. No mention was made by Pryor that they had access to any maps, but he had managed to make a compass by magnetizing two sewing needles with the magnet in the handle of his razor and using a hollow shirt stud as the pivot. Their plan involved them walking the estimated seventy kilometres (forty-five miles) to Hamburg, taking two or three nights to complete this section of their journey, and then boarding a train to the port of Lübeck where they would hope to find a Swedish ship in the harbour and smuggle themselves on board. It is known that MI9 produced a map of the port of Lübeck, similar to that of Danzig illustrated earlier in the chapter. Whilst the only copy of this map which has been discovered to date has been the *Bulletin* version, as late as June 1944 a new plan of Lübeck docks was being reported in the War Diary as having been produced and was being sent to the prisoner of war camps.

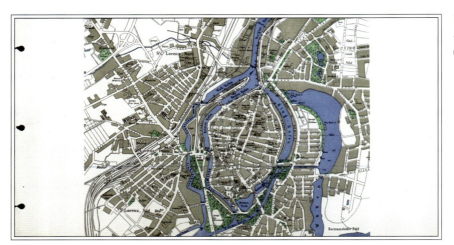

The map of Lübeck, printed on paper, from the Bulletin (with detail below).

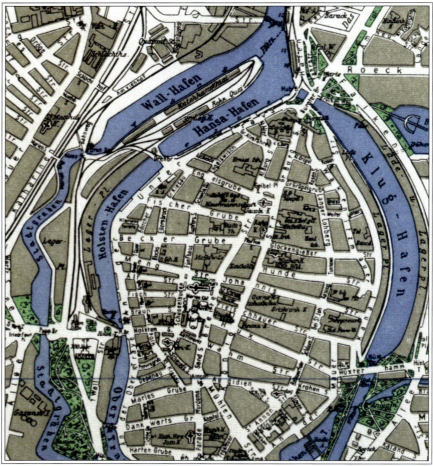

182 GREAT ESCAPES

Pryor and Wells took their opportunity on 20 September 1943. They remained on the run for two days but were sadly recaptured on the second night when they were seen by two policemen and, while Wells tried (in halting German) to talk their way out of the situation, Pryor was searched and found to have 'equipment and a jacket full of oatcakes'. They received the customary spell in solitary confinement but managed to see the positive side of their solitude as an agreeable change to the communal life they had experienced in captivity for the previous three years.

While Pryor's bid for freedom was not successful, his memoirs contain many mentions of planning escapes and he clearly recalled that escaping was the duty of every prisoner of war but with 'the whole of NW Europe under German control and with no maps or compass it seemed a pretty hopeless task'. He was involved in detail in the preparation of the escapes, often helping to dig tunnels, and helped many others to escape successfully. Pryor and his colleagues continued to plan escapes and harass their German captors as well as they could. Their activity continued up to the Normandy landings in June 1944. Later, messages were broadcast that prisoners of war should no longer attempt to escape but remain in their camps, the famous 'stay-put' order. MI9 was quick to realize that escaped prisoners of war roaming around the front line would likely be in danger and also a possible distraction to the advancing Allies. He was liberated by the Allies in 1945 and went on to serve his country in peacetime as a hydrographic surveyor in the Royal Navy.

Liberated naval and merchant seaman prisoners of war at Marlag and Milag Nord camp at Westertimke, 29 April 1945.

Lieutenant David James RNVR

The escape of Lieutenant David Pelham James of the Royal Naval Volunteer Reserve (RNVR) was notable since he did succeed in his bid for freedom via the Baltic ports' escape route. James had made multiple escape attempts throughout his captivity. His final and successful attempt was from the Marlag and Milag Nord camp on 10 February 1944, disguised as a 'distressed Swedish sailor', arriving back in the UK on 16 March 1944. His final escape journey proved to be extremely tortuous.

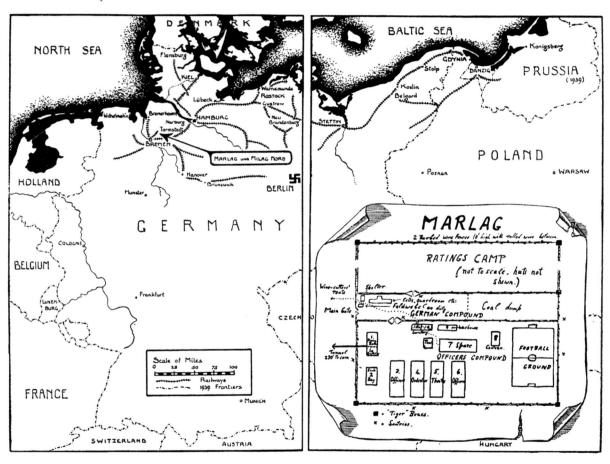

He journeyed initially by train to Lübeck, the same route which Pryor had planned to take. When he could find no Swedish ships, he travelled on to Stettin (see the *Bulletin* map of Stettin on page 172), where he similarly failed to find a suitable ship to board. Stettin was the port where MI9 had suggested in their briefings that a good place to find Swedish sailors was in the local brothel. It was

The Baltic Coast ports and a plan of Marlag camp, from David James' book, A Prisoner's Progress, published in 1947.

A sketch map of Lübeck showing where David James unsuccessfully (10 December 1943) and successfully (20 February 1944) escaped from Lübeck to Sweden, from his book, A Prisoner's Progress, *published in 1947.*

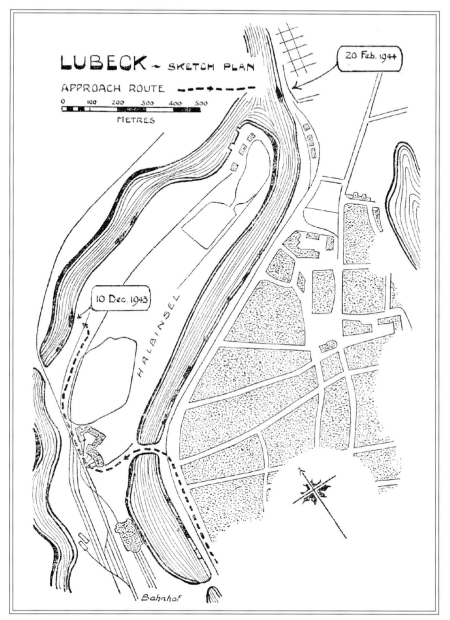

a particularly useful venue for those attempting to escape since it was reserved for foreigners: no Germans were allowed to enter. It was eventually blown up by the RAF during a raid in the summer of 1944. In desperation, as he was running out of money, James continued on to Danzig where he eventually found a Danish ship, the *Scandia*, which he boarded and on which he was befriended by a stoker. He mentioned in his account that, had he known the layout of the port and continued for another three miles, he would have found the berth of the Swedish ships. In itself, this appeared to indicate that he personally had had no access to the Danzig port plan. However, he did know enough to make his way to a port and to seek out a ship bound for a neutral country. The ship left Danzig for Denmark and docked in Lübeck port en route, at which point its destination was changed to Konigsberg. James realized that he would have to change ship. As he knew the geography of Lübeck reasonably well from a previous, failed, escape attempt, he was able to board a Finnish ship bound for Stockholm, bribed the stoker and sixty hours later he was in the safe-keeping of the British consul in Stockholm.

Flight Lieutenant Oliver Philpot

Another example of the successful use of this route was the escape of Flight Lieutenant Oliver Philpot who flew with the RAF Coastal Command out of Leuchars in Fife on 11 December 1941. His mission was to patrol the coast of Norway, taking aerial photographs. On seeing a German convoy, he attacked on a 'mast high bombing run'. The enemy returned fire and the aircraft took direct hits in the starboard engine and tail. He was forced to land in the sea and all the crew made it into the dinghy before the aircraft sank. They drifted for two days and were subsequently picked up by a German convoy. Philpot was eventually imprisoned in Stalag Luft III at Sagan. He escaped on 29 October 1943 through a tunnel and travelled by train from Sagan, over 160 kilometres (100 miles) south of Berlin, north to the Baltic ports. He made for the port of Danzig and indicated in his escape report that he was familiar with the local geography from 'the flimsy of which I had a copy and which proved invaluable'. The 'flimsy' was almost certainly sheet A4 of the port of Danzig which had been produced on both silk and tissue by MI9. He eventually boarded the Swedish ship *Bjorn* which was loading coal. Although the Captain was not happy with his presence on board, he was hidden in the coal bunker by the crew and landed in Södertälje on 3 November. He travelled by train to Stockholm and arrived at the British Legation on 4 November. Yet again, there is evidence that copies of the large-scale port plans were reaching the camps and being put to very good use by those planning to escape.

There are numerous examples, in both the published literature and contemporary records, of escapes, both successful and unsuccessful, via Swedish or Finnish ships from the Baltic ports. Certainly, Lieutenant John Pryor's unsuccessful escape with Johnny Wells from Marlag and Milag Nord camp in

1943 was an attempt to reach Lübeck. While they were captured before they had succeeded in reaching the railway, Pryor did acknowledge that Lübeck was their planned destination. James also headed directly for the Baltic ports, trying them in turn until he found a suitable ship. Similarly, Philpot made directly for Danzig, indicating that he had access to a map which had afforded him knowledge of the local geography of the port. It is a matter of record that this was the most successful escape route which MI9 established out of occupied Europe, to Sweden and eventual freedom. It is also a matter of record that MI9 produced large-scale plans of the Baltic ports and persuasive evidence has been found to support the contention that those same maps were sent to the camps for use in escapes. Yet again, MI9 planned a particular route to freedom, briefed in detail on it, and produced the maps to ensure that the prisoners of war had the very best chance of success when using this particular route.

Oliver Philpot's Stalag Luft III (Sagan) identity tag was concealed behind a photograph of his first wife Natalie, so that he could prove he was a prisoner of war and not a spy in case of his capture.

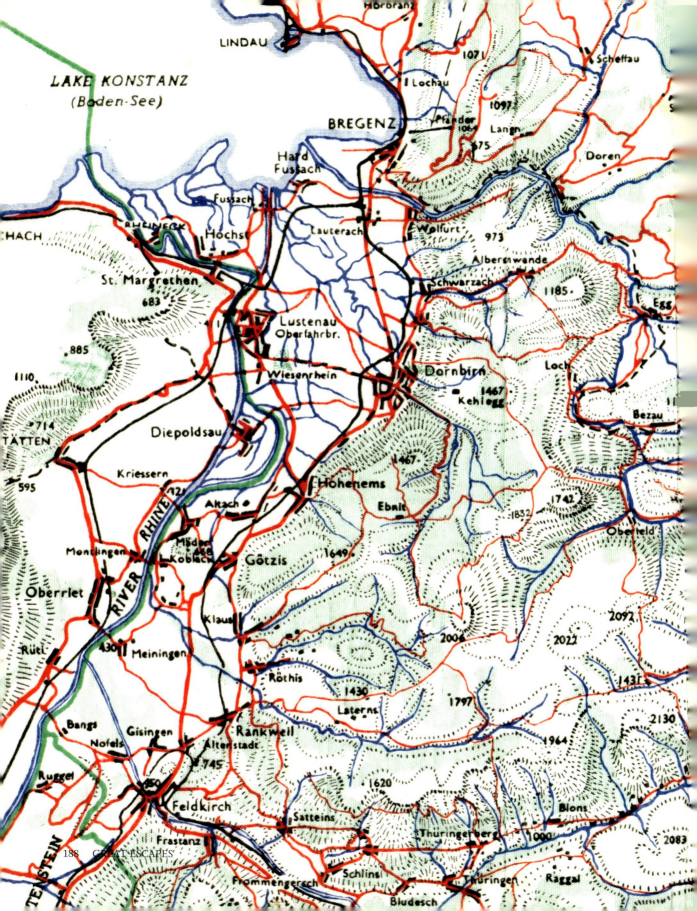

8

COPYING MAPS IN THE CAMPS

'For some a map is not a piece of paper: it is a passport . . .'
(from a TV advertisement, 2003, collected by Dr David Forrest)

Prisoners of war spent much time and energy copying maps from originals which MI9 succeeded in getting into the camps. Multiple copies of the maps needed to be produced when multiple escapes were planned and executed. If only one map got through, there was a clear need to copy it and the most straightforward way was simply to trace a copy. Airey Neave described this practice in *They Have Their Exits*: 'less technical minds studied languages, copied maps and collected stolen articles of civilian clothing'. However, to identify such copies in a British map collection is exceptional. The discovery of two such maps afforded the opportunity to look closely at the copying techniques used by the prisoners of war and also to try to identify the MI9 maps from which they had been drawn. If the maps could be identified, it would provide some independent evidence of the success of MI9's escape and evasion mapping programme since their very existence bears witness to the fact that they must have been used in successful escapes: how else could they have found their way to the UK?

COPYING MAPS

When the individual camp histories came to be written at the end of the war, they often contained useful check lists of the sort of escape aids which were needed and found to be most useful. Certainly, that for Stalag Luft III (Sagan) contained a very detailed chapter on the type of escape materials which had been needed. Maps at varying scales were described, such as large-scale port plans 'showing location of quays used by neutral shipping' and 'maps of neutral frontiers'. There was also mention of the need for a map tracer and enlarger. Interestingly, in the Colditz camp history, it is clear that the prisoners were able to steal such a machine, known as a pantograph, from the canteen and put it to very good use, although it was not made clear why such a machine would be in the canteen in the first place. Use of the pantograph not only meant that maps could be traced and enlarged at the same time, but also that multiple copies could be produced through repeated use of the machine.

Detail from sheet A6 covering the eastern end of Lake Constance (see Appendix 1). It covers part of the area shown in the manuscript map on page 192. This escape and evasion map also included two photographs.

PRINTING MAPS

Some prisoners of war were able to set up miniature printing works. Wallis Heath, who had been an officer in the Royal Corps of Engineers, recounted the story of how the prisoners of war in Oflag XI (Braunschweig), initially hand copied maps which they had received from MI9. However, he and Philip Evans, a prisoner in the same camp, had experience of printing and so they decided to set themselves up as the Brunswick Printers. They did so in the knowledge that the Allies were advancing and the camp apparently planned a mass escape as they feared German reprisals.

British prisoners of war welcome the arrival of liberating American troops at Oflag XI in Braunschweig (Brunswick), 12 April 1945.

Showing considerable ingenuity, they were apparently able to create both the substitute artwork and printing plates they required. They removed some limestone tiles from the toilet area and ground them clean to ensure the surface would be suitable for use as lithographic stones to serve as printing plates. They traced the detail of the maps on to the tiles using carbon paper which, one can only assume, they had stolen or acquired through bribery from their German guards. Separate reverse tracings were made for each colour on the map, black for text and railways, red for roads and blue for rivers. These were to act as the individual printing plates for each layer of detail on the map. They then went over the linework with a fine pen dipped in boiled margarine. In essence, they were using the lithographic principle that water and oil do not mix and utilizing what was essentially the collotype method of reproduction. Tiles were coated with gelatine, taken from the tins of meat in Red Cross food parcels, to act as a sensitizing agent. Inks were made from all sorts of powders, sometimes even those intended as stage make-up. Those elements of map detail drawn in the margarine retained the coloured ink. They constructed a printing press from wooden floor boards covered in leather with a roller made from a window bar. The tiles were used in turn to build up the map image. While the maps made by the Brunswick Printers were never apparently used as the Braunschweig camp was liberated peacefully, there is some limited evidence that similar techniques were employed elsewhere and the resultant maps used successfully by escaping prisoners of war.

TWO MANUSCRIPT ESCAPE MAPS

The first manuscript map shown over the page, was clearly drawn in Oflag VB (Biberach). That deduction is based on the fact that Oflag VB was located in Biberach which is shown at the northern edge of this map. The map extends from this point to the Swiss frontier, which was the escape route which would logically be taken from that particular camp, since it represented the shortest distance to freedom. A number of high profile escapes took place from Biberach. While many of them, not least that of Hugh Woollatt, Neave's companion on the journey back from Switzerland (see photograph on page 164), did escape via the Schaffhausen Salient, some successfully crossed the Swiss border at the eastern end of Lake Constance. Indeed, the Germans realized the success of that route very early on and guarded that particular area so closely, that eventually most escapers from Biberach chose to travel the longer western route via Schaffhausen.

The map is in manuscript form and had been hanging on a library wall for many years before being properly identified. The local record simply stated that the map had been deposited in the collection in 1947: the donor was not named but the map was described as a World War II prisoner of war map. It was relatively straightforward to identify the MI9 map from which the Biberach tracing had been made: sheet L32-2/Konstanz of GSGS 3982 [Fabric]. The detail is so closely related, as seen in an extract of that map shown beside the manuscript map below, that it is clearly identifiable as the source map.

The second manuscript map shown on page 194 was far less easy to identify since it was so badly faded. However, the use of software, by staff in the GeoMapping Unit of the School of Geography in Plymouth University, to enhance the image has enabled the original map to be redrawn (as shown on page 195) and the MI9 source to be identified. It would appear that the

Manuscript map of route from Biberach to the Swiss frontier and detail from GSGS 3982 [Fabric] sheet L32-2/Konstanz covering a similar area.

source map was sheet A, produced by Waddington at MI9's direction from a Bartholomew map (refer to Appendix 1 and Chapter 3 for details of the source map). The detail from sheet A shown below covers a similar area to the original map. Sheet A was certainly one of, if not the, earliest map in the whole of MI9's map production programme, witnessed by the map being identified by the first letter of the alphabet. It is also possibly the map which MI9 termed 'Double Eagle', providing small-scale coverage of both Germany and Austria, and adjacent areas.

The remarkable chance survival of both these manuscript maps, albeit unsupported by surviving written testimony regarding the escape attempts concerned, demonstrates a critically important aspect of MI9's escape programme, namely the copying of maps that were smuggled into the prisoner of war camps. Serving as unique testimony to the efforts of both the prisoners of war and MI9, they provide the final example in this book to demonstrate the extent to which the maps could, and did, provide a veritable passport to freedom for some prisoners of war.

RIGHT:
Detail from the southern section of Sheet A in the Bartholomew series, showing southern Germany and the surrounding countries. It was used as the source for the manuscript escape map reproduced over the page.

OVERLEAF:
Manuscript escape map and a digitally enhanced image of the same, identifying features in the area near the Swiss border and thus proving that it was originally copied from Sheet A in the Bartholomew series.

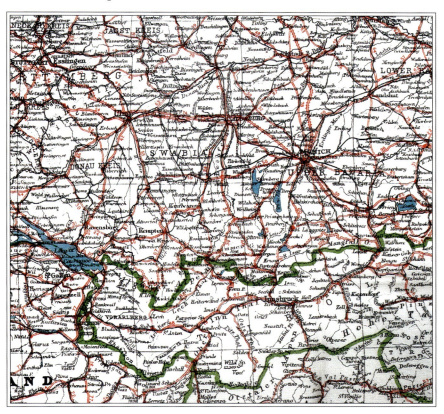

COPYING MAPS IN THE CAMPS

PRIME MINISTER'S
PERSONAL MINUTE

10, Downing Street,
Whitehall.

SERIAL No. M.788

Message from the Prime Minister to the Prisoners of War.

In this great struggle in which we are engaged, my thoughts are often with you who have had the misfortune to fall into the hands of the Nazi.

Your lot is a hard one, but it will help you to keep your courage up to know that all is well at home. Never has the country been so completely united in its determination to exterminate Nazidom and re-establish freedom in the world. Our strength grows daily, and assistance flows from America in ever-increasing volume. In high-hearted confidence we press forward steadily along the road to certain victory.

Keep yourselves fit in mind and body, so that you may the better serve our land, and, when peace comes, play your part in establishing a happier, safer homeland.

God bless you all.

Winston S. Churchill

August 3, 1941.

9

MI9 AND ITS CONTRIBUTION TO MILITARY MAPPING

'It was too little, too late.'
(*Fight Another Day*, by J. M. Langley)

The work of MI9 has been afforded little recognition in the study of the history of British intelligence organizations of World War II. Hitherto, the MI9 mapping programme has been given even less attention and yet, as this study has shown, it amounts to a very significant chapter in the evolution of British military mapping and a notable episode in the history of twentieth century cartography. While there has been a long history of military mapping on silk and other fabrics, it transpires that MI9, for all sorts of reasons, was never in a position to benefit from that history and experience. Nonetheless, it was able to mount a very significant programme of escape and evasion mapping despite the role played by Christopher Clayton Hutton, arguably a self-possessed, eccentric and narrowly focused individual who was allowed to operate with little or no restriction. The context in which the branch was operating also took its toll. As a newly spawned intelligence branch, it lacked supporters from the beginning and was, too often, regarded as a threat by its sister organizations in the intelligence arena. To what extent, therefore, was MI9 able to fulfil its role as the organization responsible for escape and evasion despite the challenges which it undoubtedly faced: in essence, was it really 'too little, too late' as Langley wrote?

THE SIGNIFICANCE OF MI9'S MAPPING PROGRAMME

W. G. V. Balchin was commissioned by the Royal Geographical Society (RGS) in 1985 to record the role played by geographers in World War II. He considered the maxim that 'geography had always been vital to the prosecution of war' and the extent to which it was 'the intelligent use of geographical knowledge that outwits the enemy and wins wars'. The value of training in geography had been realized during World War I. Geography was deemed vital to the successful prosecution of war in three ways, namely in intelligence, logistics and action. This awareness had resulted in the rapid development of the subject in universities and schools during the inter-war period.

It is an interesting aspect of Balchin's findings that at no point did he mention MI9 or its escape and evasion mapping programme in his report

Winston Churchill was a supporter of MI9's work. His personal minute to prisoners of war was smuggled into the camps in cigars or using invisible inks printed onto cotton handkerchiefs.

to the RGS. And yet, he must have been aware of it since Foot and Langley's definitive account of the organization is included in the list of references accompanying his report. He certainly considered the role of maps and even mentioned the value of terrain analysis (an indication of the extent to which the area offered opportunities for concealment and obstacles to movement). The evolution of European topographic mapping services was mentioned, as were the responsibilities of the Geographical Section General Staff (GSGS) in the War Office: even the role of geographers in the Special Operations Executive (SOE) is covered in detail. Of MI9 and its mapping programme, however, there is absolutely no mention. Since Balchin relied heavily on interviewing those geographers who were still alive and ready to recount their experiences, the single reason for that omission can only be the exclusion of any geographers from direct involvement in MI9's mapping programme until the war was well progressed. The reason for that, in turn, was arguably attributable to Hutton's single-mindedness in the way in which he drove his section's programme and simply never communicated with those in the War Office who could have offered detailed technical and, not least, cartographic production support.

Peter Collier, in describing the development of air survey techniques and their use in producing and updating mapping in World War II, wrote that 'Ultimately, it was the Allied capacity for the mass production of maps, together with the weapons of war, that was to prove decisive.' What he wrote, however, was arguably as applicable to MI9 escape and evasion maps as it was to the operational coverage produced by D.Survey and it is this lacuna in the historiography of cartography in World War II which this study has sought to bridge.

It was envisaged from the beginning of this study that a key outcome of the research would be the creation of as full an inventory as possible of all the escape and evasion maps that were produced between 1939 and 1945. Discovering the sheer size and scope of the MI9 mapping programme was, however, never anticipated. MI9's escape and evasion maps were produced as practical, problem-solving items and they represented a marked change in military attitudes to capture and captivity in enemy hands. Access to a map that depicted the area through which they must travel to reach a safe haven after escaping incarceration was quite literally the prisoner's 'get out of jail' card: as revealed earlier, many of the maps were despatched inside Monopoly boards or in playing cards. While Hutton cannot be regarded, at least initially, as cartographically literate, those officers who received the maps, copied them and used them successfully as the prime escape aid, most certainly were. Unlike

Hutton, they belonged to a new generation which had been educated between the wars and had apparently benefited from the upsurge in the teaching of geography, a direct result of the hard-won experience of World War I. For many of them, reading maps and navigating routes with the aid of maps was second nature.

HUTTON'S ROLE

The mapping programme was an essential aspect of MI9's entire endeavours and Hutton was the key personality involved directly and was the initiator of the mapping programme. Hutton proved to be a very enigmatic personality. He was variously described by those who knew and worked with him as both an eccentric and a genius. In reviewing the various members of Norman Crockatt's team, Foot and Langley described Hutton as 'the joker in the pack', wayward yet original, with an apparently limitless supply of both enthusiasm and ingenuity, possessing no regard for either rules or officialdom, and was apparently left by Crockatt very much to his own devices.

Hutton appeared to have had little awareness of the extent to which MI4/GSGS/D.Survey could help in his mapping endeavours and there is little doubt that MI4's absence from London at a critical time may be part of the explanation.

The role of Christopher Clayton Hutton was instrumental in MI9's mapping programme. Here he is surrounded by some of his special creations, in a photograph taken from the back cover of Official Secret, *published in 1960.*

The fact that the department responsible for the cartographic revision of operational mapping using aerial reconnaissance was distant from the centrality of military awareness and operations between 1939 and 1942 undoubtedly took its toll: quite how much is open to discussion and interpretation.

Hutton, apparently oblivious of the importance of that absence from the equation, overcame the cartographic challenges presented, without any real awareness of the true nature of those challenges. Evidently, without ever realizing that organizations existed within Government (indeed, in the same Ministry) which could have rendered his job very much easier, Hutton nevertheless managed to identify individuals and companies who could progress his strategy. Even so, by excluding the operational mapping organization from the start and by being unaware of commercial companies with considerable expertise in map production and printing, Hutton certainly created challenges for himself and his organization that need never have existed. Time-consuming activities and significant costs were undoubtedly incurred which could have been avoided or, at least, contained. It can, therefore, be argued that Hutton succeeded with the mapping programme despite his own lack of awareness of those organizations which might have helped him more promptly, more immediately and, arguably, at less cost.

Hutton after MI9

Taken ill during the war, it seems that Hutton left MI9 on health grounds and there is some later indication that he suffered a mental breakdown. A letter he wrote from his hospital bed in February 1943 to Victor Watson, the managing director of Waddington, contained a rather strange request. He asked that he be sent a list of all the 'pretty pictures' (his coded phraseology for maps) and the amounts which had been paid to Waddington for the work. The letter ended:

> . . . with my kindest regards and many thanks for the troubles you all took to help us over a very high stile . . . I am a broken spirit at being here and doubt if I shall ever be back but cannot tell and don't much care.

Watson realized the inappropriateness of Hutton's request and responded that he could not meet the request, but was happy to send a Waddington game. In a further letter, Hutton acknowledged his misguided behaviour and apologized. Hutton had left MI9 but appeared to recover sufficiently from his illness by February 1944 to apply to SOE for possible employment. The vetting form in the file noted his previous employment with MI9, but was stamped 'NOT TO BE EMPLOYED' owing to 'traces'. The precise nature of the concerns was not recorded, possibly his health and state of mind, but his attempt to return to the intelligence world, unsurprisingly, met with failure.

After the war, Hutton ran into further difficulty with the authorities when he tried to publish his own story about MI9 and also embark on a lecture tour of the United States of America. He was keen to highlight the role he had played in MI9's escape and evasion programme and earn some money at the same time. He described how he approached the War Office in 1950 for permission to lecture and write a book about his experiences, at the same time providing proof that there was already information about the escape aids in the public domain. Indeed, some of the surplus maps had been sold off soon after the war, an undertaking which had been openly described and discussed in the press at the time. He completed a first draft of the book under the title *A Journey Has Been Arranged* and invited Air Marshall Sir Basil Embry to read and comment on it. Sir Basil offered to write the preface and in it he indicated not just his personal indebtedness to Hutton, but he also commented, 'Some people may think he is eccentric; I think he is a genius.' That completed draft was sent to his publisher, before being sent to the War Office for permission to publish, which was not forthcoming.

There was a strange twist to the story in the mid-1950s with the publication of Charles Connell's book *The Hidden Catch*. It was submitted by the publisher, Elek Books, to the Air Ministry for clearance to publish on 11 August 1955. It is clear from the files that this was the same book in essence as that which had been stopped in 1951. It is very obviously Hutton's story, although he is described throughout the book as Mr X. Permission to publish was initially refused and Hutton was warned that to ignore the decision would make him liable to prosecution under the Official Secrets Act. However, the publisher agreed to make changes to the text requested by the Air Ministry, and this resulted in permission to publish being given. The book was published in 1955. For unexplained reasons, Hutton did not realize until after publication that one of the textual changes was, in his opinion, defamatory of himself and he pursued a claim for libel against his own publisher. In May 1957, the Treasury Solicitor was informed that the action had been settled out of court and a discontinuance notice had been served.

The fractious relationship between the faceless men in the War Office, Hutton and publishers both in the UK and the USA appeared to continue and the situation was not apparently resolved to Hutton's satisfaction until the eventual publication of *Official Secret* in 1960, which appears to be a condensed form of the original draft which Hutton had written some ten years previously.

It is not clear just how much of Hutton's reticence to include much detail about the mapping programme in his book resulted from the security vetting process or from an understandable reluctance, in the light of his previous experience, to continue battling with the War Office Committee responsible for vetting publications. Some four years later in 1964, however, he did author a short article for the American journal *Popular Science* which served to highlight some of the detail of the escape gadgetry, compasses, clothing and currency, and included a section on the maps. It added nothing to the minimal detail about the maps which he had already included in his own book.

The final chapter of *Official Secret* described the various personal and professional challenges that he encountered. He appeared to have an almost obsessive personality which led him to pursue his objective with single-minded focus and total self-belief, and it may well have been the complexities of his personality that caused his eventual illness and hospitalization, and the end of his career with MI9. He certainly caused considerably more difficulties and costs than would have been encountered had he sought to involve the military map-makers from the start. It is, however, the case that all who knew Hutton, and those who wrote about him, appeared to agree that he was the right man for the job. For all his undoubted weaknesses, it appears to be the case that Hutton, virtually single-handedly, managed to mount and sustain the detailed map production programme, and the fact that he managed to do so without the help of those best placed to make his life easier, makes his feat the more impressive.

A NEW AND INEXPERIENCED INTELLIGENCE ORGANIZATION

Before considering just how successful MI9 can be judged to have been, the organization needs to be set in the context of the wartime crisis and against the backdrop of individual and departmental rivalries. Petty rivalries and jealousies apparently persisted despite the popular assumption that they were held in abeyance during the war. The newly formed organization took time to find its feet. In its early days, this was due largely to a lack of staff. It took time to recruit and staff the various sections, especially as the country was already at war and was soon sustaining heavy losses and defeats. As Head of the fledgling service, Crockatt was also aware from the beginning that, as Foot and Langley described it:

> . . . secret and semi-secret services like to work in a dense fog of security, in which the germs of inter-secret-service jealousy breed fast.

In 1939 the Foreign Office had wanted MI9 to be established under its direct control. The reason for that is not stated in the records of the time but is very likely to have been so that the Secret Intelligence Service (SIS) could exert direct control and influence on its work from the beginning. As it was, the Foreign Office's opposition was overruled and, throughout its existence, MI9 was an integral part of the War Office, where it met the needs of all three Services. It became clear from the post-war published works of those who were directly involved with MI9 at the time, especially Neave and Langley, that SIS sought to undermine Crockatt at every opportunity, and to control much of his Branch's work.

MI9 and SIS

Sir Stewart Menzies had been appointed Chief of SIS, a post generally referred to as C (the Chief's code name, after Sir George Mansfield Smith-Cumming, the first director of what would later become SIS, who always initialled papers he read, C). Menzies' appointment followed the death of Admiral Quex Sinclair in November 1939 and occurred barely a month prior to the creation of MI9 and Crockatt's appointment. Menzies' Assistant Chief was Colonel (later Sir) Claude Dansey. Dansey was older (sixty-three years old at the outbreak of war) and it was sometimes felt that Menzies deferred to him. Dansey was regarded as, and apparently preferred to be, something of an *eminence grise*, rather than the man in charge. As Foot and Langley put it 'He could have broken Crockatt, or anyone else in MI9 . . . and Crockatt knew it'. Certainly Crockatt kept his organization relatively small and tight, and appeared to avoid any direct confrontation with SIS, even when he was aware of the absence of its support from his section's work. Crockatt was, however, very clear about SIS's role and thinking, and the extent to which they were motivated by their history and experience during World War I. At the end of the war, he wrote:

> The oft repeated statement that Nurse Edith Cavell, who apparently worked for SIS during the last war, had been discovered through assisting a prisoner of war seemed to dictate the whole attitude of SIS towards the Section. They were determined to prevent escapers and evaders from involving them in any way. This attitude may have been correct from their own security aspect, but it was a terrific handicap to those trying to build up an organisation.

Crockatt sought at every stage not to cross SIS's path. He encouraged his staff to 'take it quietly and remember we are playing very much a lone hand'. SIS had clearly not forgotten that in October 1915 Edith Cavell had been shot

by the Germans for hiding British soldiers in her clinic in Brussels and helping them to escape to neutral territory. Whilst Crockatt understood their approach, he clearly regarded it as detrimental to the work of MI9. He was supported in that view by both Langley (who worked for SIS but within MI9) and by Neave (who worked alongside Langley in MI9). SIS never realized that an increase in support for MI9 was needed rather than the negative approach which Crockatt described:

> Their rather negative form of support continued to the last and had the inevitable effect of restricting the scope of the Section's work in every country with which it was concerned.

Neave expressed his view very clearly in the second book he wrote, *Saturday at MI9*. To him, the cynical belief of the other intelligence branches appeared to be that airmen shot down by the Germans were a matter of minor importance. He strongly disagreed, describing their actions as 'this negative campaign'. He was convinced that MI9 was simply not taken seriously enough, that it lacked influence and was afforded the lowest possible priority with the Air Ministry. He also clearly disliked Dansey's behaviour towards MI9 and spoke of the 'battles with Uncle Claude'.

MI9 had no direct representation in the War Cabinet, whereas SIS did. Menzies knew by July 1940, well before Crockatt did, of the intention to create SOE. Menzies resented the creation of this subversive organization, independent of SIS's control. Together with Dansey, he viewed the establishment of SOE as undermining their long-held monopoly over the control of undercover work in enemy territory. They supposed that SOE would likely subvert, or at least hinder, their work of obtaining intelligence. It is very probable that this was the unstated reason for Menzies and Dansey's offer to Crockatt in August 1940 to set up an escape route for MI9 from Marseilles into Spain. Crockatt accepted the offer and the details of the new organization were arranged by Dansey. He chose Donald Darling, code name 'Sunday', to establish the escape route to run from Marseilles to Barcelona and Madrid, and then via Lisbon or Gibraltar to London. Darling's cover was as Vice-Consul, responsible for refugees and he was based in Gibraltar. Dansey also placed Jimmy Langley, who had been recruited to SIS after his successful escape, in MI9 as the main interface between the two organizations.

Langley, later joined by Neave, was potentially SIS's Trojan horse inside MI9, although it never quite worked out in that way. Langley later recalled that he needed courage to face up to Dansey as he always reduced him to 'a petrified

Execution of Edith Louisa Cavell (1865-1915), British nurse and humanitarian, sentenced to death for helping Allied soldiers escape from German-occupied Belgium during World War I. An imaginative illustration from the French newspaper Le Petit Journal, 7 November 1915. It was thought that Cavell also worked for SIS, who felt that she was betrayed because of her work helping soldiers escape, a fate SIS wished to avoid for their agents in World War II.

jelly'. That was quite an admission from a man who had escaped from a hospital in Dunkirk with a suppurating amputation wound and had successfully escaped through France to Marseilles and then back to London. Langley also recognized that, 'as a late arrival in the intelligence community . . . escape and evasion was very much at the tail end of queue'.

SIS, effectively, had control of MI9's escape lines in Western Europe and how they operated. However, Dansey was determined that SIS's valuable agents should never be used on what he regarded as 'a thoroughly unproductive clandestine pastime' and that nothing should ever be allowed to interfere with SIS's work to collect intelligence information 'from all possible sources, by every feasible means, the world over'. The fact that MI9 was also producing valuable intelligence must have rankled with SIS and was certainly not something which SIS was ever ready openly to acknowledge.

When F. H. Hinsley and others wrote *British Intelligence in the Second World War* in a ten-year period from 1979, it was written very much from SIS's perspective, and made very little mention of MI9, its role or contribution, throughout the five volumes. Although Hinsley and his collaborators did make some limited mention of SIS's relationship with SOE, there is no mention of SIS's view of MI9, and yet, as has been shown, MI9 was a source of extremely valuable intelligence and could get answers to some vexing operational questions. SIS was on standard distribution for the receipt of all such reports emanating from MI9, and yet that fact is not even hinted at in Hinsley's 'official' and very detailed review of the story of British intelligence in World War II. It was as if there was a conscious decision to write the fledgling escape and evasion organization out of the history books. The record was certainly corrected by Foot and Langley when they wrote the book which has come to be regarded as the definitive history of MI9 and in which Crockatt's personal leadership role and the challenges he faced are openly discussed and acknowledged. Their book was published in the same year that the first volume of Hinsley's history appeared.

MI9 and the Air Ministry

SIS was not, however, the only branch of Government service which wanted to control MI9. In 1943 the Air Ministry proposed that Crockatt be replaced by a senior RAF officer and that all MI9's responsibilities should be placed under their jurisdiction. It was rumoured that the debate went to the highest levels of the War Cabinet and that Churchill himself ruled that MI9 should remain under War Office control. The fact that there were competing players seeking to control MI9 probably made it doubly difficult for SIS ever to mount a successful takeover. It must also have made life for Crockatt far from comfortable, working daily in the knowledge that others were seeking to exert influence and control over his every move.

MI9 and SOE

Contacts with other secret departments, such as SOE, were arguably not made early enough, although a useful and mutually beneficial relationship was eventually created. Neave provided very specific evidence of this when he described MI9's support to Operation Frankton (popularly known as the Cockleshell Heroes), SOE's daring Commando raid on German shipping on

Bill Sparks of the Royal Marines, c.1943. He was one of the survivors of Operation Frankton, an SOE raid on shipping in the German occupied French port of Bordeaux in December 1942. MI9 provided special maps, compasses and escape route details.

the River Gironde in Bordeaux in December 1942. He indicated that, after the success of the raid, two of the survivors made for Ruffec and the escape route which had been pre-planned with MI9:

> With the aid of special maps and compasses with which they had been supplied by MI9 they continued marching until, at dawn on December 18th, weak and hungry, they reached Ruffec. They had walked nearly a hundred miles.

Internecine intelligence disputes

In his own end-of-war summary, Crockatt indicated his own awareness of the sensitivity which attached to MI9's relationship with SIS. He knew that SIS senior officers hated SOE since the turmoil and unrest fomented by SOE made life for SIS agents in area 'awkward'. Whether or not SIS was aware of MI9's support for SOE is not known, but had they been, it would likely have confirmed their worst fears. Similarly escapers, evaders and their local helpers were 'anathema to SIS officers of the old school'. Potentially MI9 might disrupt their work of obtaining intelligence. The fact that the simmering internecine hostilities which existed between SIS and MI9 never actually broke out into open warfare was almost certainly due to Crockatt's low-key behaviour.

There appeared to be no structures in place, however, to minimize the tribalism and perhaps frictions were inevitable. It is certain that personalities also played a part, as highlighted by both Neave and Langley in the case of Dansey. In the Conclusion to his unpublished thesis, 'Beset by Secrecy and Beleaguered by Rivals', Thomas Keene made some very pertinent remarks about the lamentable frictions which existed between SOE and SIS. He has also highlighted the extent to which the clash of interests and lack of clarity in the demarcation between the two organizations was never resolved, as it should have been. The same situation was also commented on by Leo Marks who indicated, in *Between Silk and Cyanide*, the extent to which SIS resented SOE since it threatened its monopoly: he described the mutual antipathy as having 'the growth potential of an obsession'. Those same views can be applied to SIS and MI9 as precisely the same situation undoubtedly existed between those two organizations. Since there was never any overall control of the two organizations by a single Minister of Cabinet rank, only Churchill could have resolved the matter. It is clear that Churchill, an escaper himself during the Boer War, had a particular interest in the work of MI9, reflected in the personal letter sent to the prisoners of war, hidden inside cigars, a letter clearly designed to help sustain morale in the camps. The extent to which the Prime Minister's personal support, in the same way he supported the role of Bletchley Park, could have been Crockatt's saviour is not known and does not appear to be reflected in the records. It might have made Crockatt's task easier, had the Prime Minister sought instead to ensure that inter-departmental conflict was stopped by clearer means of demarcation. In the final analysis, the various secret organizations should have been demonstrating a vested, joint and coordinated national interest in winning the war, rather than fighting internal battles.

THE CLOSURE OF MI9

It is perhaps indicative of the personal toll which the six years of leading MI9 took on Crockatt that he chose to retire to private life very soon after the cessation of hostilities and left the winding down of MI9 to Sam Derry, the organizer of one of the escape lines (from Italy). However, as a way of keeping in touch with those who had worked in MI9 or MI19, Crockatt did establish the 919 dining club, in much the same way that SOE established the Special Forces Club. Sadly, Crockatt's club folded soon after his death in 1956, whereas SOE's club survives in London to this day.

Despite detailed and lengthy searches through the records, the date of the formal closure of MI9 has not been identified, although there is every indication that it started as early as July 1945 and its officers were demobilized in 1946. Certainly, the Branch remained active long enough to ensure that prisoners of war returning from the camps in both Europe and Asia completed short questionnaires in an attempt to learn lessons about escape experiences which might prove valuable in any future war and to identify any collaborators for possible prosecution. It also organized a lengthy exercise to get food aid to those who had helped on the escape routes and to identify those who should receive honours for the support they had given to escapers and evaders, recognizing the extent to which they had placed their own lives in mortal peril. The task was eventually completed by the Air Ministry and they decided to fund the RAF Escaping Society, which continued in existence into the 1990s, ensuring that those who been directly involved with MI9, their families, dependents and descendants, received the support they needed long after the war had ended.

DID MI9 MEET ITS REMIT?

Any judgement of MI9's success, or failure, needs to be set against this background in which it was forced to operate. In seeking to make a judgement, it is fair to be reminded of the principal objectives with which they had been charged in 1939 and which have been described in detail in Chapter 1. They fell under three broad headings, namely morale of prisoners of war, the acquisition of intelligence, and escape and evasion.

In terms of morale, the all-pervading philosophy of escape-mindedness which Crockatt and his staff tried to inculcate seemed to have improved the morale of the prisoners of war. The camp histories make much of the extent to which the involvement of the camps in the activity of planning escapes contributed to the maintenance of a mood of optimism. The time spent in planning and executing escapes kept the prisoners of war occupied through the many days, months and, in some cases, years of captivity. They felt that, despite their captivity, they were still actively contributing to the war effort. Many prisoners of war were aware that their camp was in covert contact with the War Office, although how and when communications occurred was known only to the relatively few coded letter writers or to those using the radios. News sent via the coded letters and, later on in the war, via the wireless contacts, encouraged them in their endeavours. They received news of what was happening, especially after the Allies had landed in Normandy, and started to push back the German front.

Those at home had their morale boosted with the return of the successful escapers and evaders. This was expressed by Neave in his Introduction to Langley's book, *Fight Another Day:*

> It was the sudden reappearance of airmen reported lost, at RAF Stations, that had so deep an impact. When the great raids on Germany began, and losses began to mount, these miraculous returns from the unknown encouraged the whole RAF. They knew that, even if wounded, they had a chance of avoiding capture. The lectures and escape aids of MI9 increased their confidence. More than once we took an airman back to his Station . . . the joy with which he was greeted made all our efforts worth while.

Similarly, the families of the captured military personnel, both officers and other ranks, were alerted by MI9 to the extent to which their sons were able to continue to contribute to the war effort despite their incarceration behind barbed wire.

Another key responsibility given to MI9 from the beginning was to collect information from British prisoners of war through maintaining contact with them during captivity and after successful repatriation, and disseminate the intelligence obtained to all three Services and appropriate Government Departments. Perhaps one of the most surprising aspects of the research behind this book has been the discovery of the extent to which MI9 was able to establish coded contact with the camps and use it, not only as a crucial link in supporting escape activity, but also as the means to encourage the camps to provide whatever intelligence they could to support the war effort. The coded letter system was often used to seek answers to some very specific enquiries about Allied losses and the state of the German war machine. The camps proved to be a remarkable fifth column and one which, to date, has received little or no recognition.

In terms of the escape and evasion activity itself, the sheer output of the production of escape aids, leaving aside the mapping programme for the moment, was impressive. Between 1 January 1942 and 25 August 1945, MI9 arranged for the production and despatch of 423,075 escape packs, 275,407 purses, over 1,700,000 compasses and related devices, and 434,104 miscellaneous items, including such items as gramophone records and other leisure items, all previously described in Chapter 4. The list is impressive enough and becomes even more so if the whole of the escape and evasion mapping programme is added to it.

THE SUCCESS OF MI9'S MAPPING PROGRAMME

As a result of this research into the escape and evasion mapping programme, it is now possible, for the first time, to reveal an estimate of the numbers of escape and evasion maps which MI9 actually produced. A conservative figure would be that 243 individual items were produced and in excess of one and three quarter million copies were printed (see Appendices 1–9 for full details). Over half of the copies produced were of three of the sheets in [Series 43] covering Western Europe which it is believed were printed for operational use prior to the D-Day landings. Excluding this, a cautious estimate of the total printed for escape and evasion purposes would still be in excess of three quarters of a million copies.

A critically important further finding has been the sheer scale of the covert involvement of commercial companies by MI9 in order to mount such levels of cartographic production in a wartime situation. By any measure, this appears to have been a quite remarkable effort, the more so when one takes account of the self-inflicted production challenges which resulted from Hutton's own insularity and lack of cartographic awareness, experience or training. If numbers are an indicator of success, then certainly the programme can be deemed to have been successful. While a significant proportion of the print volume took place after D.Survey became involved, and probably reflected an awareness of

Sheet 43C from [Series 43], one of the maps printed in large numbers, probably for operational use in preparation for the D-Day landings.

the value of fabric maps in an operational situation, it is by any standard an impressive record of map production.

It is also worthwhile considering the numbers of escapers and evaders who managed to return to the UK before victory was declared. Over 35,000 are reported to have returned, numbers which equate to the size of three Army divisions. The direct evidence that these escapers used escape and evasion maps is remarkably hard to find, however. None of the published sources which can be ascribed to Hutton for their information explores the mapping programme in any detail in spite of the fact that it was clearly an early priority for the newly established MI9. There are many other books which recount the detail of escape and evasion, though none appears to include detail of the maps. The escapes described range from the truly ingenious and inspirational to the bizarre and sometimes opportunistic. However, even those which describe the escape route in minute detail fail to describe the maps used or their source. Most mention maps in passing and highlight the extent to which they figured on the 'must-have' list of most successful escapes. Time and again, escapers include mention of the provision of money and maps by the escape committees in the oflags and stalags as part of their aids to escape but offer little or no supporting detail. Why should that be? In discussion with Professor Foot in January 2012, over sixty-five years after the end of World War II, he was surprisingly still unwilling to explain how he had managed to keep hidden the MI9 map of France and Spain with which he had been issued prior to being dropped as an SAS officer in occupied France in 1944. When prompted, he explained that all MI9's pre-operational briefings had always included the warning that the maps should never be discussed, and he insisted that he had always abided by that, although he did imply that he had succeeded in keeping his map from the Germans even when he had been captured, by hiding it inside the lining of his jacket. The map had remained in his possession and he produced it during the discussion.

EPITAPH

Despite the scale of activity, Langley indicated that the only epitaph he could find for the work of Crockatt, Neave, Darling and himself was that 'it was too little, too late'. While arguably he should have known from his own experience just how true, or not, that description was, there is much to indicate that his judgement was far from balanced. Perhaps he was too close to be objective and dispassionate. As an escaper himself, he identified very closely with the fate of those who were required to spend months and often years of their young lives as guests of the Third Reich. His judgement was probably also coloured by his understandable desire to help as many as possible to freedom. He confessed that he blamed no-one for the shortcomings of MI9 and certainly not 'Uncle Claude' and SIS. He felt that the fault, such as it was, lay 'in the inability of anyone to see that the apparently impossible was possible'. For Langley, it was the lack of foresight and even of imagination, which resulted in the failure to grasp the considerable potential of organized escape and evasion which he personally regarded as so self-evident. He believed that, had MI9 been granted the same freedoms and status as SOE, much more could have been achieved. It is clear that he experienced anger and frustration at what he regarded as MI9's failures, particularly the loss of life on the escape lines of those members of the French, Belgian and Dutch resistance who were captured and executed by the Gestapo or who died in the concentration camps.

Crockatt himself wrote an interesting epitaph of his own in the historical record of MI9's work. His short section on 'lessons learned' spoke volumes of his own view about the task MI9 had been set and the extent to which its personnel had had to deal with unexpected obstacles. He set out the ways in which any future escape and evasion organization might benefit from MI9's experience with the very particular recommendation that 'an adequate staff for planning purposes of obtaining information likely to help escapers and evaders is allowed at the outset'. It was acknowledged by all three Services and the USA that MI9 had made 'an important contribution to the war effort'. Many owed their life and liberty to equipment devised and issued by Section Z, but it was also acknowledged in the record that experts with their own workshop to devise and make new equipment would have proved a valuable addition to the organization.

The research behind this book has uncovered not only an immensely significant episode in the history of British military mapping, but has also shed light on a hitherto largely unacknowledged aspect of British intelligence activity in World War II. It has revealed for the first time a remarkable chapter in the production, use and application of military mapping on silk (and other fabrics) and, moreover, it has shown the extent to which the maps contributed to the mission of aiding escapers and evaders upon which MI9 embarked in 1939. To make this contribution to the history of twentieth century cartography, it has been necessary to look closely at MI9 and its role, and to understand the dynamics of escape from prisoner of war camps. It is for others to progress the story through wars which have occurred since 1945 and possibly through the systematic exploration of the extent to which the Germans were aware of, and sought to counteract, the programme.

In the final analysis, the manner in which MI9 executed its escape and evasion mapping programme, for the benefit of those thousands of men captured and imprisoned by Hitler's regime, has proved to be a quite outstanding cartographic feat and one which needed to be told. It has proved to be one of the unrecognized triumphs both of this nation's military mapping history and its military intelligence capability operating in a wartime scenario. Who knows what further triumphs might have been accomplished had Crockatt and his team been allowed to proceed unhindered by inter-departmental politics, adequately staffed and funded, and directly supported by the undoubted experience and expertise of the nation's military mapping organization.

APPENDICES

Appendices 1–9	Maps Known to have been Produced by MI9: A Carto-Bibliography	
	Introduction	216
Appendix 1	**Fabric maps based on Bartholomew originals (and maps with similar numbering system): extant copies identified**	218
Appendix 2	**Fabric maps presumed to be based on Bartholomew originals: no extant copies identified**	226
Appendix 3	**Europe Air 1:500,000 GSGS 3982 [Fabric]**	227
Appendix 4	**Norway 1:100,000 GSGS 4090 [Fabric]**	232
Appendix 5	**[Series 43]**	234
Appendix 6	**[Series 44]**	235
Appendix 7	**[Series FGS]**	237
Appendix 8	**Miscellaneous maps**	238
Appendix 9	**Maps produced for the *Bulletin***	240
Appendix 10	**Decoding a hidden message**	242

MAPS KNOWN TO HAVE BEEN PRODUCED BY MI9: A CARTO-BIBLIOGRAPHY

INTRODUCTION

A detailed description of each of the various escape and evasion series/groups of maps produced by MI9 in the period 1939–45 is given in Chapter 3. The following appendices have been compiled to identify the individual sheets in each series. Each appendix lists the sheets identified from the various historical records investigated and from libraries and museums around the country which have been visited in the course of the research. Bibliographic details are provided for each sheet. Production details are included wherever possible and the various collections holding extant copies of the sheets are identified. Brief and tailored introductions are provided at the beginning of each appendix.

In all, a total of 243 individual maps have been identified and extant copies of over 170 of them have been located in seven libraries and museums and four private collections. It is estimated that in excess of 1,860,000 copies were printed in the five years from the early months of 1940 until production was halted in 1945. The total number of printed copies was clearly boosted by the large numbers of [Series 43] (in excess of one million) which were apparently printed for operational rather than escape and evasion purposes in the run-up to D-Day. Notwithstanding that, it is clear that in excess of three quarters of a million copies of the MI9 maps were apparently produced for escape and evasion purposes.

PRINCIPAL MAP REPOSITORIES

The first stage in compiling this information was to identify the principal libraries and museums in which escape and evasion maps were to be found. The UK's record set of these maps was held in the Defence Geographic Centre (DGC) but is presently (2015) being deposited in The National Archives (TNA). Access was granted to DGC on the basis of a Freedom of Information request. This set of the maps was the most comprehensive set available anywhere, and yet it was incomplete. This collection does, however, form the core of the detailed map inventory or carto-bibliography that follows.

The next most significant collection in terms of its size is that to be found in the Royal Air Force Museum at Hendon. The British Library Map Library holds part of the Waddington company archive. This company was enlisted by MI9 as the first printers of the silk maps. The company also manufactured many of the items (board games and playing cards) used to smuggle the maps into the prisoner of war camps. Smaller, but no less important, collections of the maps were found in the Intelligence Corps Museum, the Macclesfield Silk Museum, the Cumberland Pencil Museum and in a number of private collections. Sadly the collection housed in the Imperial War Museum was not available for public access due to a lengthy re-cataloguing and refurbishment programme. However, an online check of their holdings revealed no map which was not held elsewhere.

KEY TO LOCATION OF EXTANT COPIES IN THE TABLES THAT FOLLOW

1   Defence Geographic Centre (DGC): this is the UK's record collection of military maps and is destined for deposit in The National Archives
2   The National Archives (TNA)
3   British Library Map Library (BL)
4   Intelligence Corps Museum (ICM)
5   Royal Air Force (RAF) Museum
6   Macclesfield Silk Museum
7   Cumberland Pencil Museum
8   Private collections (for reasons of confidentiality and security, no detail about private collections is disclosed)

PREPARING THE CARTO-BIBLIOGRAPHY

The lack of standard cartographic information provided a challenge in the compilation of a carto-bibliography. It is also worth reflecting on the current state of carto-bibliographic practice. Anglo-American Cataloguing Rules Edition 2 (AACR2) are in the process of being superseded by Resource Description and Access (RDA) but this is in the very early stages of implementation in the USA and is even further behind in the UK. More importantly, it does not, as yet, contain guidance for maps, although some map librarians in North America are understood to be working on what will eventually be the new version of *Cartographic Materials: a Manual of Interpretation for AACR2*. In an attempt to ensure that the form of the carto-bibliography presented here will be of value to professional map librarians and archivists, the instructions in the Manual have been adhered to in so far as is possible and, additionally, advice has been sought from map librarians/archivists in both The National Archives and the British Library. It has been important to ensure that the data fields described are capable of being captured into an automated map catalogue system in the future.

The solution to the challenges posed by MI9's lack of adherence to usual cartographic identification procedures has been to utilize the standard cartographic technique of showing in square brackets [ ] any information which does not appear on the printed maps and which helps to identify the maps. The series number and series title, for example [Series 43] and Series GSGS 3982 [Fabric], have been rendered in this form to aid identification. The orthography of the original maps is retained in the bibliographical entries which follow.

PRINT DATES

Some of the differences between the D.Survey card index and the Waddington list (see Chapter 3) relate to the print dates ascribed to the maps. This was potentially a significant research issue since it was important to construct a carto-bibliography which contained print volume figures which are as accurate as possible. Where the dates ascribed to print dates are only days apart, these have been regarded as the same task, with the Waddington date probably representing the placing of the order by the Ministry of Supply, on behalf of MI9, and the War Office date probably representing the receipt of the stock from the Waddington company. Where the dates ascribed are months apart, these have been regarded as separate tasks and the print volumes are listed as separate tasks. While every effort has been made to resolve the differences, this has not proved possible in all cases.

PRINT MEDIA

Wherever possible, the print medium of each map has been described. This information has been derived either from the existence of an extant copy (silk, tissue or man-made fibre/MMF) or from the records where the medium is usually listed as paper or fabric. Occasionally, the records divulged further information on the nature of the paper being used: the abbreviations used were ML (Mulberry Leaf paper), MLS (Mulberry Leaf Substitute) and RL (Rag Lithographic paper).

APPENDIX 1

Fabric maps based on Bartholomew originals (and maps with similar numbering system): extant copies identified

The maps listed in Appendix 1 are an amalgam of those escape and evasion maps which are based directly on the maps of John Bartholomew & Son Ltd of Edinburgh, together with some maps which utilized a similar numbering system but which are clearly not based on Bartholomew maps. They have been combined in one appendix for ease of reference. Part 1 of the appendix contains a detailed geographical description of each map. The entries are tabulated under the column headings: sheet number, geographical co-ordinates, date, geographical coverage, scale, dimensions (length × width), detail (of colours), and notes. The few dates which appear have been taken from the date of the boundary information in the map legend: in general, however, the escape and evasion maps are not dated and do not carry their own production details.

Part 2 of the appendix contains details of the production information. The first column shows the sheet number and indicates whether it was produced singly or in combination with another sheet. The second column indicates the medium on which the map was printed. The first part of the entry denotes the actual medium of the extant copies: tissue, silk or MMF (man-made fibre). The second part of the entry denotes the information derived from the print record: paper or fabric. The third column indicates the size of the print run and the fourth column indicates the print or despatch date. In all cases, this detail has been derived from the print record. The final column indicates the location of extant copies of the sheet.

Part 1: geographical description of the maps

Sheet number	Geographical co-ordinates	Date	Geographical coverage	Scale	Dimension (length × width)	Detail
A	5°40'–17°+E 46°30'–55°+N		Germany, Poland, Austria, Czechoslovakia, and parts of Switzerland, Italy, Yugoslavia, Denmark, Sweden, Holland, Belgium and France	1:2M	516 mm × 479 mm	Black detail, red roads, green international boundaries; some copies have blue water plate [1]
A1			German–Swiss frontier area, section of the Schaffhausen Salient in the region of Schleitheim	2.5 in. = 5 miles (c.1:126,720)	250 mm × 190 mm	Black detail, red boundary and pylons, green woods, blue water
A2			Denmark and Baltic seaboard S to Swiss border, Holland/Belgium/France borders, and parts of Austria, Poland and Czechoslovakia	1:4M	255 mm × 255 mm	3 colours: black, red, grey/green
A4			Port of Danzig at mouth of River Vistula	(c.1:16,000)		Black detail and blue water plate [2]
A6		June 1943	German–Swiss frontier and eastern end of Lake Konstanz, mouth of Rhine and Liechtenstein	3 in. = 10 miles (c.1:210,000)	380 mm × 310 mm	Black detail, blue water, red roads, green boundary and woods [3]
BL2 [4]	6°–7°50'E 50°35'–51°40'N		SE Holland, E Belgium and adjacent parts of Germany	1:600,000	215 mm × 255 mm	3 colours: black, red, grey/green
BL3	7°40'–9°55'E 50°40'–51°45'N		Area of Germany centred on Kassel	1:600,000	215 mm × 255 mm	3 colours: black, red, grey/green
C	+4°W–8°+E 47°30'–54°+N		France north of Loire, Luxembourg, Belgium, Holland and W Germany: no detail over GB except extreme SE	1:2M	435 mm × 537 mm	Black detail, red roads, green international boundaries
(as C above but no sheet number)			Sheet is almost identical to C above, but detail is shown over GB to N and W extent of sheet			
D	3°W–8°+E 42°–47°+N		France south of the Loire and east to Switzerland	1:2M	434 mm × 523 mm	4 colours: black detail, red roads, blue lakes, yellow international boundaries
F	–4°–36°+E –60°–70°+N	1939 and 1940	Northern part of Scandinavia Inset of North Sea Connections	1:3M (also produced at 1:2M with smaller geographical coverage)	485 mm × 589 mm	3 colours: black, red, green [5]
G	–6°–30°+E –54°– 60°+N	1939 and 1940	Southern part of Scandinavia and the Baltic Sea	1:3M (and 1:2M)	485 mm × 589 mm	3 colours: black, red, green [6]
H	+9°W–3°+E –35°–42°+N		SW France, Portugal, most of Spain, part of Balearics and adjacent N Africa Inset of Balearics and Canary Islands	1:2,571,000	504 mm × 578 mm	3 colours: black, red, green.
H2	+8°W–9°+E –36°–44°+N		Iberian peninsula, S France, NW Italy, Corsica, Sardinia, Balearics and adjacent part of N Africa	1:3M	399 mm × 599 mm	3 colours: black, red, green

Sheet number	Geographical co-ordinates	Date	Geographical coverage	Scale	Dimension (length × width)	Detail
J3	7°–15°+E 40°–46°+N		N Italy Larger scale inset of Swiss frontier and inset of Rome	1:1,378,000	628 mm × 521 mm	4 colours: red, blue, black, green Not Bartholomew specification Based on native Italian maps Italian, French and English language legend [7]
J4	13°–18°+E 37°–42°+N		S Italy Same scale inset of Sicily	1:1,378,000	394 mm × 482 mm	
J5	14°30'–16°15'E 39°–40°10'N		NW part of Sardinia	1:275,000	517 mm × 570 mm	Not Bartholomew specification Full topography with contour plate
J6	8°6'–9°52'E 40°–44°10'N		Sardinia, Gulf of Polesto and area S of Naples	1:275,000	517 mm × 570 mm	English language legend
J7	2°15'–2°55'E 37°45'– 38°10'+N [8]		Sicily, area of Mt Etna	1:110,000	510 mm × 615 mm	Not Bartholomew specification Full topography with contour plate Red, black, blue colours English legend
J8	1°28'–2°15'E 37°40'– 38°05'+N [8]		Sicily	1:110,000	510 mm × 615 mm	
K	19°–33°E 28°–36°N		Cyrenaica and Crete (no coverage of Cyprus)	1:1,657,000	502 mm × 587 mm	3 colours: black, red, grey/green
K1	details identical in all respects to K above					
K2	–0°–24°+E 12°–37°+N		Partial overlap with K/K1, extending W and S	c.1:6M	491 mm × 551 mm	3 colours: black, red, green.
K3	+20°W–4°+E 12°–36°+N		Morocco, Algeria, Tunisia, Italian Libya and Spanish Rio do Oro. Partial overlap with K2, extending W to Atlantic Ocean with Canary and Madeira islands	1:6M	526 mm × 469 mm	3 colours: black, red, green
K4	+15°W–5°+E –0°–15°+N		Partial overlap with K3, extending S to Gulf of Guinea	1:5M	526 mm × 465 mm	3 colours: black, red, green
K5	–0°–15°+E –0°–20°+N		Partial N/S overlap with K2, extending S and partial E/W overlap with K4, extending E	1:5M	593 mm × 465 mm	3 colours: black, red, brown [9]
K6	8°–28°+E +8°S–6°+N		Overlaps with K5, extending S	1:4M	593 mm × 465 mm	3 colours: black, red, grey/green
ME/01/F	–34°–54°+E 28°–42°N		E Mediterranean E to Iran and N Persian Gulf N to Black and Caspian Seas	1:4,500,000	443 mm × 505 mm	6 colours: black, red, green, orange, blue, purple [10]
MP1	96°–110°E –2°–20°N		Laos, Cambodia, S Thailand, Malay Peninsula and N Sumatra	1:4M	638 mm× 513 mm	3 colours: black, red, green
N	–36°–43°+E 12°–18°+N		Eritrea extending E across Red Sea to Yemen	1:2M	465 mm × 471 mm	2 colours: black, red
O	–35°–42°E 4°–12°N		Abyssinia	1:2M	550 mm × 466 mm	2 colours: black, red
P	42°–51°+E 4°–12°N		Somaliland	1:2M	476 mm × 583 mm	2 colours: black, red
Q	–38°–47°+E 2°S–3°+N		Kenya Colony and Juba River	1:2M	476 mm × 583 mm	2 colours: black, red

Sheet number	Geographical co-ordinates	Date	Geographical coverage	Scale	Dimension (length × width)	Detail
R	20°–31°E −39°–48°+N	1938–1940	Balkans and adjacent areas	1:2M	607 mm × 454 mm	3 colours: black, red, green
R1	identical in all respects to R above					
R2	−26°–38°+E −46°–60°+N		W USSR and adjacent parts of E Europe N to Leningrad and Gulf of Finland	1:3M	589 mm × 501 mm	3 colours: black, red, green
R3		1939–1940	3 maps: 1. Denmark and S Sweden 2. Baltic States, Poland, Czechoslovakia, adjacent areas 3. N Norway, N Sweden, Finland, part of USSR, Baltic States	1:3M	589 mm × 501 mm	3 colours: black, red, green
S1	14°–22°+E 40°–48°+N	1938–1940	S Italy, NW Greece, Albania, Austria, Yugoslavia, Hungary, and parts of Rumania, Bulgaria, Czechoslovakia	1:2M	602 mm × 460 mm	3 colours: black, red, green
S2	19°–29°+E 35°–42°+N	1938–1940	Greece, Albania and parts of Yugoslavia, Bulgaria, Turkey Inset of Crete at same scale	1:1,750,000	530 mm × 622 mm	3 colours: black, red, green
S3	10°–30°+E 36°–48°+N		Sicily, Italy, Greece, Crete, Turkey, Bulgaria, Yugoslavia, Albania and parts of Rumania, Austria, Hungary, Switzerland	1:3M	530 mm × 622 mm	3 colours: black, red, grey/green
T1	26°–36°+E 34°–48°+N		W part of Black Sea, S to Asiatic Turkey and Cyprus, and parts of Rumania, Sea of Marmara, E Aegean and E Crete	1:3M	605 mm × 455 mm	3 colours: black, red, green
T2	42°–60°E 38°–48°+N		Caspian Sea and surrounding area	1:3,200,000		
T3	28°–46°+E 28°–38°+N		E Mediterranean, east to cover Syria and Iraq, and parts of Egypt and Cyprus	1:3M	463 mm × 611 mm	3 colours: black, red, green
T4	44°–62°E 28°–38°+N		Iran, N of Bushire and adjacent border areas	1:3M	463 mm × 611 mm	3 colours: black, red, green
T7	no known details					
T8	no known details					
U6	no known details					
U7	no known details					
U8	no known details					
W	−32°–44°+E 28°–38°+N		Egypt, Cyprus, Asiatic Turkey, Syria, Transjordan, Iraq, and parts of Iran and Arabia	1:3M	533 mm × 600 mm	3 colours: black, red, green
W1	identical in all respects to W above					
W2	−46°–62°+E −26°–38°+N		Arabia, Iraq, Persian Gulf, Iran, Afghanistan, SW USSR	1:3,750,000	518 mm × 585 mm	3 colours: black, red, grey/green
X			Map from Spittal in Austria to Mojstrana in Jugoslavia Inset of whole route from Oflag VIIC in Salzburg to Mojstrana	1:100,000 1:1,350,000	560 mm × 524 mm	4 colours: black, red, green, blue [11]

Sheet number	Geographical co-ordinates	Date	Geographical coverage	Scale	Dimension (length × width)	Detail
Y			Map of Schaffhausen Canton with goings information to aid PoWs escaping to Switzerland from Germany	1:100,000	430 mm × 590 mm	
Y2			2 maps: 1. Schaffhausen and S Germany 2. S Germany	1:350,000		3 colours
Y3	no known details					
Z	12°–19°+E 53°–57°+N		N Germany, N Poland, Bornholm, E Denmark, S Sweden	1:1M	470 mm × 504 mm	3 colours
General Map of Ireland (no sheet number shown)			N area on one side and S area on reverse	1:633,600	440 mm × 552 mm	4 colours: blue, black, red and orange for boundaries of 6 Ulster counties
Zones of France (no sheet number shown)			France including cross Channel area but SE England not shown Inset of Zones shown in NW	[circa 1:2M]	620 mm × 580 mm	6 colours [12]

1. Bartholomew print number 311 appears in SW corner of some copies. Some copies have no sheet number
2. Known also to exist at larger scale with no sheet number
3. Contains two ground photos and a sketch map
4. Also marked B2 in top margin
5. Bartholomew print number A40 appears in NW corner of some copies
6. Bartholomew print number B39 in SW corner of some copies
7. J3 and J4 also known to exist with geographical areas reversed, scale reduced to 1:1,500,000 and different insets, i.e. J3 with insets of Sicilia and Sardegna; J4 with inset of part of Corsica
8. Based on Rome meridian, 12°27'7.1" East of Greenwich
9. Relief shown by hachures
10. Relief shown by layers
11. Contains detailed goings information to guide escapers across border into Jugoslavia
12. Note same title also used at Appendix 8. Relief shown by layers

Total = 59 sheets (15 of which are not Bartholomew's specification)

Part 2: production details and location of extant copies

Produced singly/ combination	Print medium		Print runs	Print/despatch date	Location of extant copies
A	tissue				1, 5
	silk	fabric	10,000	15.7.42	1, 5*
A1	tissue				5
A2	tissue				7
A4	silk	fabric	150	5.10.42	4
		paper	150	5.10.42	
A6	tissue				5
A/C		fabric	20,000	30.6.42	
		fabric	550	1.12.42	
		fabric	1,000	21.12.42	
	tissue				5

Produced singly/combination	Print medium		Print runs	Print/despatch date	Location of extant copies
A/G					
BL2	tissue				7
BL3	tissue				7
C	tissue				3, 5
	silk	fabric	10,000	8.7.42	5
C/D	MMF		25,000	21.8.42	5
C/H2			10,000	21.8.42	
D	tissue		10,000	17/20.7.42	5
D/H2	MMF		7,500	24.7.42	1, 4, 5
			10,000	21.8.42	
			25,000	20.5.43	
			10,000	21.6.43	
			10,000	6.7.43	
			5,000	29.7.43	
			5,000	9.8.43	
F	tissue				3
F/G	MMF	fabric	5,000	7.10.42	1, 5
H	tissue				1, 5
	silk				1
H2	tissue	paper/fabric	2,000	15.7.42	1, 5
H2/K2	MMF	fabric	2,500	12.11.42	
H2/K3			5,000	31.7.42	
	MMF	fabric	22,330	7.9.42	5
			15,000	27.4.43	
J3	tissue	paper	2,000	7.1.42	5
	MMF	fabric	5,000	7.1.42	4
J3/J4	MMF	fabric	8,000	7.1.42	1, 5, 6
		fabric	10,000	23.6.42	
			5,000	2.6.43	
J4		fabric	5,000	7.1.42	1, 4, 5
		paper	2,000	7.1.42	
J4/K1	MMF				5
J5/J6	MMF		350	11.6.43	1, 4
J7/J8	MMF		350	8.6.43	1, 4
K					
K/W	MMF				1
K/W1	MMF				1, 5
K1	tissue	paper	1,000	23.6.42	
	MMF	fabric	1,000	23.6.42	

Produced singly/ combination	Print medium		Print runs	Print/despatch date	Location of extant copies
K1/K2	MMF	fabric	10,000	1.7.42	1, 4, 5, 6
		fabric	22,500	7.9.42	
		fabric	500	8.10.42	
			15,000	20.5.43	
K2	tissue	paper	1,000	23.6.42	1, 5, 6
	MMF	fabric	1,000	23.6.42	
K3		paper	1,000	8.7.42	
	MMF	fabric	1,000	15.7.42	6
K3/H2	MMF		15,000	27.4.43	1, 4
K3/K4	MMF				1, 4
K5/K6	MMF				1, 4, 5
K6	MMF				1
ME/01/F	silk			6.10.42	1
MP1	silk				1, 5
N	tissue	paper/fabric	1,000	24.7.42	1
	MMF	paper/fabric	1,000	24.7.42	5
N/O	MMF				5
O	tissue	paper/fabric	2,000	24.7.42	1, 5
	MMF	paper/fabric	2,000	24.7.42	1
P	tissue	paper/fabric	2,000	23.6.42	
	MMF	paper/fabric	2,000		
P/Q	MMF	fabric			1, 5
Q	tissue	paper/fabric	2,000	23.6.42	
	tissue	paper/fabric	2,000	23.6.42	
R	tissue				1
R1	tissue	paper/fabric	2,000	23.6.42	1, 5
	MMF	paper/fabric	2,000	23.6.42	
R1/S1	MMF				1, 5
R2/R3	MMF				1, 4, 5
R3	silk				4
S1	tissue	paper/fabric	2,000	23.6.42	1, 5
S2	tissue				4
S2/S3	MMF		1,500	2.9.42	1, 4, 5
			10,000	27.4.43	
T1	tissue	paper	100	29.1.43	
	MMF	fabric	45	29.1.43	
T1/T2	MMF				4
T1/T3	MMF				1, 4, 5

Produced singly/combination	Print medium		Print runs	Print/despatch date	Location of extant copies
T2		paper	1,000	23.3.42	
		fabric	1,000	23.6.42	
T2/T4					5
T3	tissue	paper	100	29.1.43	
	MMF	fabric	45	29.1.43	
T3/T4	MMF				1, 5
T4	silk	paper/fabric	145	23.6.42	4
	tissue				
	MMF				
T7		fabric	50	22.1.43	
T8		fabric	50	22.1.43	
U6		fabric	50	22.1.43	
U7		fabric	50	22.1.43	
U8		fabric	50	22.1.43	
W	silk				1, 5
W1/W2	MMF				1, 5
W2	silk				1, 5
X	silk				1, 3
Y	silk				3
Y2/9Y3†	MMF				1, 5
	paper				3
Y3	RL		1,000	23.6.42	
Z	silk				1, 5
General Map of Ireland	silk				1
Zones of France	bank paper		550	14.1.42	5
	fabric		550	14.1.42	
	bank paper		550	23.1.43	

* this copy is not marked A but is identical in all other respects to those which are so marked. Assumed to be the sheet called Double Eagle in the records.
† refer to Appendix 8 for details of this map.

Total number of copies produced = 348,570

APPENDIX 2
Fabric maps presumed to be based on Bartholomew originals: no extant copies identified

These are maps which are assumed to have been produced based on Bartholomew maps but for which neither extant copies nor related records have ever been found. They represent either gaps in the assumed sequential numbering system, sheets which are mentioned on adjacent sheet diagrams or, in one case, a sheet which was seen at an Antiques Fair but details were not noted. In cases where the sheet number has the prefix 9 followed by an alphabet letter, it could be that these are identical or similar to sheets which carry simply the alphabet letter and are listed at Appendix 1. It has not, however, been possible to prove this hypothesis to date.

Sheet number	Known details
E	Europe 1/2M
H3	none
J	believed to cover Italy
L1	none *
M	covers Darfur
T7	none
T8	none
U6	none
U7	none
U8	none
Y3	none
9B	1:2,350,000 Germany†
9C	1:2,350,000 France†
9D	1:2,350,000 France (SECRET) †
9F/G	20 miles to 1 in. = 1:1,267,200 Norway/Sweden†
9H	20 miles to 1 in. = 1:1,267,200 Spain/Portugal†

* reported as sighted at map fair but no other details noted
† probably minor variants of sheets listed under only an alphabet letter at Appendix 1

Total = 16 sheets

APPENDIX 3
Europe Air 1:500,000 GSGS 3982 [Fabric]

Escape and evasion versions were produced of seventy-four sheets in the Europe Air, 1:250,000 scale, series GSGS 3982. They were, however, reduced to 1:500,000 scale and were, therefore only one quarter the sheet size of the original series: as such, they were often referred to as 'miniatures' or 'handkerchief maps'. With the exception of the scale factor, no other details were changed. As a result, the font size of place and feature names appears very small, although still legible, and the detail is dense. The sheets were produced sometimes singly and sometimes in combination, but it has not proved possible to identify the various combinations. Additionally, one sheet, N33/9, was produced at 1:375,000 scale. Four irregular sheets were also produced at the scale of 1:420,000 in a block centred on Arnhem ('Dutch Girl'): these do not carry sheet numbers but are rather marked as sections 1–4. They are listed in a separate table at the end of this appendix.

Where extant copies of the maps have been discovered, these have been used to confirm sheet-lines. The sheet numbers and titles have been extracted initially from the print record and spellings, including the use of diacritic marks, have been confirmed against extant sheets and/or the map index.

Sheet number	Sheet title	Geographical extent	Compiled and drawn in the War Office	Print runs	Print dates	Location of extant copies
K32/3	Florence	10°–12°E/43°–44°N		45 fabric + 100 paper	20.10.42	
				100 fabric (1 side only)	6.5.43	
K33/1	Ancona	12°–14°E/43°–44°N		45 Fabric + 100 paper	9.10.42	
				100 fabric (1 side only)	6.5.43	
K33/4	Terni	12°–14°E/42°–43°N		45 fabric + 100 paper	16.10.42	
				100 fabric (1 side only)	6.7.43	
K33/8	Fóggia	14°–16°E/41°–42°N		45 fabric + 100 paper	16.10.42	
				100 fabric (1 side only)	6.5.43	
K33/11	Naples	14°–16°E/40°–41°N		45 fabric + 100 paper	20.10.42	3
				100 fabric (1 side only)	6.5.43	
K33/12	Táranto	16°–18°E/40°–41°N		45 fabric + 100 paper	20.10.42	
				100 fabric (1 side only)	6.7.43	
L32/1	Basle	6°–8°E/47°–48°N		45 fabric + 100 paper	10.7.42	4, 8
				100 tissue	13.10.43	
L32/2	Konstanz	8°–10°E/47°–48°N		45 fabric + 100 paper	24.7.42	8
				100 paper	24.3.43	
				100 tissue	13.10.43	
				500 paper	1.3.44	
L32/3	Innsbruck	10°–12°E/47°–48°N		45 fabric + 100 paper	21.8.42	8
				100 tissue	13.10.43	
				500 paper	1.3.44	

Sheet number	Sheet title	Geographical extent	Compiled and drawn in the War Office	Print runs	Print dates	Location of extant copies
L32/5	St Gotthard	8°–10°E/46°–47°N		45 fabric + 100 paper	21.7.42	3, 4, 5, 8
				100 fabric (1 side only)	6.5.43	
				50 paper	27.7.43	
				500 fabric	4.10.43	
				500 paper	1.3.44	
L32/6	Bolzano	10°–12°E/46°–47°N	1940	45 fabric + 100 paper	24.10.42	4, 5, 8
				100 paper	30.4.43	
					6.5.43	
					4.10.43	
					1.3.44	
L32/7	Turin	6°–8°E/45°–46°N	1939		21.10.42	4, 5, 8
				100 fabric (1 side only)	6.7.43	
				500 fabric	4.10.43	
L32/8	Milan	8–10°E/45–46°N	1938	45 fabric + 100 paper	21.10.42	4, 8
				100 fabric (1 side only)	6.5.43	
				50 paper	27.7.43	
				500 fabric	4.10.43	
L32/9	Verona	10°–12°E/45°–46°N		550 fabric	22.10.43	
L32/10	Cúneo	6°–8°E/44°–45°N	1939	45 fabric + 100 paper	21.10.42	4, 5, 8
				100 fabric (1 side only)	6.5.43	
				500 fabric	4.10.43	
L32/11	Genoa	8°–10°E/44°–45°N	1939	45 fabric + 100 paper	9.10.42	4
				100 fabric (1 side only)	6.5.43	
				500 fabric	4.10.43	
L32/12	Bologna	10°–12°E/44°–45°N		550 fabric	22.10.43	
L33/1	Salzburg	12–14°E/47–48°N		45 fabric + 100 paper	25.9.42	
				100 tissue	13.10.43	
				500 paper	1.3.44	
L33/2	Graz	14°–16°E/47°–48°N		500 MLS tissue	7/8.44	
L33/3	Sopron	16–18°E/47°–48°N		500 MLS tissue	7/8.44	3
L33/4	Údine	12°–14°E/46°–47°N	1939	45 fabric + 100 paper	30.9.42	4
				100 fabric (1 side only)	6.7.43	
				550 fabric	15.10.43	
				500 paper	1.3.44	
L33/5	Klagenfurt	14°–16°E/46°–47°N	1939	45 fabric + 100 paper	6.10.42	4
				100 tissue	13.10.43	
L33/7	Venice	12°–14°E/45°–46°N		100 fabric (1 side only)	29.4.43	
				550 fabric	20.10.43	

Sheet number	Sheet title	Geographical extent	Compiled and drawn in the War Office	Print runs	Print dates	Location of extant copies
L33/10	Ravenna	12°–14°E/44°–45°N		550 fabric	22.10.43	
M31/3	Antwerp	4°–6°E/51°–52°N		100 fabric + 200 paper	29.10.42	
M31/6	Brussels	4°–6°E/50°–51°N		45 fabric + 100 paper	10.7.42	
				100 paper	24.3.43	
				100 tissue	13.10.43	
M32/1	Essen	6°–8°E/51°–52°N	1936	45 fabric + 100 paper	10.7.42	4, 8
				100 paper	8/9.9.43	
				100 tissue	13.10.43	
				200 MLS	8.1.44	
M32/2	Kassel	8°–10°E/51°–52°N	1936	45 fabric + 100 paper	31.7.42	4, 8
				100 tissue	13.10.43	
				200 MLS	8.1.44	
M32/3	Halle	10°–12°E/51°–52°N		45 fabric + 100 paper	21.8.42	
				100 tissue	13.10.43	
				500 paper	1.3.44	
M32/4	Cologne	6°–8°E/50°–51°N	1936	45 fabric + 100 paper	10.7.42	4, 8
				100 paper	30.4.43	
				100 fabric	4.10.43	
				200 MLS	8.1.44	
M32/5	Frankfurt	8°–10°E/50°–51°N	1936	100 fabric + 200 paper	29.10.42	
				100 tissue	13.10.43	
M32/6	Erfurt	10°–12°E/50°–51°N		500 ML tissue	3.44	
M32/7	Saarbrücken	6°–8°E/49°–50°N		45 fabric + 100 paper	10.7.42	4, 8
				100 tissue	13.10.43	
				200 paper	8.1.44	
M32/8	Mannheim	8°–10°E/49°–50°N		500 ML tissue	3.44	
M32/9	Nürnberg	10°–12°E/49°–50°N		500 ML tissue	3.44	
M32/10	Strasbourg	6°–8°E/48°–49°N	1937	45 fabric + 100 paper	10.7.42	4, 8
				100 tissue	13.10.43	
M32/11	Stuttgart	8°–10°E/48°–49°N		500 ML tissue	3.44	
M32/12	Munich	10°–12°E/48°–49°N		45 fabric + 100 paper	29.12.42	
				100 tissue	13.10.43	
M33/1	Leipzig	12–14°E/51°–52°N		45 fabric + 100 paper	31.7.42	8
				500 paper	7.9.42	
				100 tissue	13.10.43	
M33/2	Görlitz	14°–16°E/51°–52°N		150 fabric (1 side only)	29.10.42	1
				150 paper	29.10.42	
				100 tissue	13.10.43	

Sheet number	Sheet title	Geographical extent	Compiled and drawn in the War Office	Print runs	Print dates	Location of extant copies
M33/3	Breslau	16°–18°E/51°–52°N		45 fabric + 100 paper	25.9.42	
				100 tissue	13.19.43	
				500 paper	1.3.44	
M33/4	Chemnitz	12°–14°E/50°–51°N		45 fabric + 100 paper	25.9.42	
				100 tissue	13.10.43	
M33/5	Prague	14°–16°E/50°–51°N		45 fabric + 100 paper	11.8.42	8
				100 tissue	13.10.43	
M33/6	Schweidnitz	16°–18°E/50°–51°N		45 fabric + 100 paper	25.9.42	
				100 tissue	13.10.43	
M33/7	Plzeň	12°–14°E/49°–50°N		500 M of S tissue	7/8.44	
M33/8	Jihlava	14°–16°E/49°–50°N		500 M of S tissue	7/8.44	
M33/9	Brno	16°–18°E/49°–50°N		500 M of S tissue	7/8.44	
M33/10	Passau	12°–14°E/48°–49°N		500 M of S tissue	7/8.44	
M33/11	Linz	14°–16°E/48°–49°N		500 M of S tissue	7/8.44	3
M33/12	Vienna	16–18°E/48–49°N		500 M of S tissue	7/8.44	
M34/1	Łódź	18°–20°E/51°–52°N	1941	45 fabric + 100 paper	14.8.42	4, 8
				100 tissue	13.10.43	
M34/4	Gleiwitz	18°–20°E/50°–51°N		500 ML tissue	3.44	
N32/5	Flensburg	8°–10°E/54°–55°N		500 ML tissue	3.44	3
N32/6	Kiel	10°–12°E/54°–55°N		500 ML tissue	3.44	3
N32/7	Groningen	6°–8°E/53°–54°N	1936	45 fabric + 100 paper	11.8.42	4, 8
				100 tissue	13.10.43	
N32/8	Hamburg	8°–10°E/53°–54°N	1936	45 fabric + 100 paper	21.8.42	4, 8
				100 tissue	13.10.43	
N32/9	Lübeck	10°–12°E/53°–54°N		100 fabric + 200 paper	22.10.42	
				50 paper	27.7.43	
				100 tissue	13.10.43	
				200 tissue MLS	8.1.44	
N32/10	Osnabruck	6°–8°E/52°–53°N	1936	100 fabric + 200 paper	22.10.42	4
				100 tissue	13.10.43	
N32/11	Hanover	8°–10°E/52°–53°N		500 ML tissue	3.44	
N32/12	Magdeburg	10°–12°E/52°–53°N		500 ML tissue	3.44	
N33/4	Stralsund	12°–14°E/54°–55°N		45 fabric + 100 paper	20.9.42	
				100 tissue	13.10.43	
N33/5	Kolberg	14°–16°E/54°–55°N		45 fabric + 100 paper	30.9.42	
				100 tissue	13.10.43	
N33/6	Stolp	16°–18°E/54°–55°N		45 fabric + 100 paper	30.9.42	
				100 tissue	13.10.43	
N33/7	Neustrelitz	12°–14°E/53°–54°N		500 ML tissue	3.44	

Sheet number	Sheet title	Geographical extent	Compiled and drawn in the War Office	Print runs	Print dates	Location of extant copies
N33/8	Stettin	14°–16°E/53°–54°N		45 fabric + 100 paper	11.8.42	8
				1000 paper	8.10.42	
				50 paper	27.7.43	
				100 paper	8/9.9.43	
				100 tissue	13.10.43	
				500 paper	1.3.44	
N33/9*	Schneidemühl	16°–18°15'E 53°–54°10'N		tissue		3
N33/10	Berlin	12°–14°E/52°–53°N	1938	45 fabric + 100 paper	31.7.42	4, 8
				100 tissue	13.10.43	
				200 paper MLS	8.1.44	
N33/11	Landsberg	14°–16°E/52°–53°N		500 ML tissue	3.44	
N33/12	Poznań (Posen)	16°–18°E/52°–53°N		500 ML tissue	3.44	
N34/4	Danzig	18°–20°E/54°–55°N	1938	45 fabric + 100 paper	31.7.42	1, 4, 8
				100 paper	8/9.9.43	
				100 tissue	13.10.43	
				200 MLS	8.1.44	
N34/5	Konigsberg	20°–22°E/54°–55°N		500 ML tissue	3.44	
N34/7	Marienwerder	18°–20°E/53°–54°N		500 ML tissue	3.44	
N34/10	Plock	18°–20°E/52°–53°N		500 ML tissue	3.44	
N34/11	Warsaw	20°–22°E/52°–53°N		45 fabric + 100 paper	14.8.42	8
				100 tissue	13.10.43	

*sheet N33/9 is identical in specification to sheets in GSGS 3982 but does not carry the series number. Its scale is 1:375,000 and it is of marginally greater geographical extent than the other sheets in this series.
ML tissue = Mulberry Leaf paper; MLS = Mulberry Leaf Substitute; RL = Rag Litho

Total number of map sheets = 74

Total number of copies printed = 35,100

Dutch Girl: scale 1in. = 6.56 miles (approximately 1:420,000)

	Geographical extent	Print runs	Print dates	Location of extant copies
Section 1, 2, 3, 4		4,400	6.42	
1	4°5'–7°E/51°21'–52°40'N	985	2.43–8.44	4
2	4°5'–6°50'E/50°–51°21'N	794	2.43–8.44	4
3	6°40'–9°40'E/51°20'–52°40'N	1,050	2.43–8.44	4
4	6°50'–9°40'E/50°–51°20'N	992	2.43–8.44	4

Total number of copies printed = 8,221

APPENDIX 4
Norway 1:100,000 GSGS 4090 [Fabric]

Escape and evasion versions of thirty-three sheets of Norway, 1:100,000 scale, series GSGS 4090, were produced on silk. Thirty-one sheets were monochrome and two were printed in four colours. The original maps were based on the Oslo meridian and the conversion factor to Greenwich was, therefore added to the operational series. The sheets are in a block to the north of Oslo and adjacent to the Swedish border. They are all single-sided: none appears to have been produced in combination. This Appendix provides individual sheet details, sheet number, sheet title, geographical extent, date of the original Norwegian map, print date of the GSGS paper series and sheet dimensions. The final column indicates the location of the extant copies of the sheets which have been found. No details of the volume of the print runs have ever been discovered.

Sheet number	Sheet title	Geographical values (based on Oslo meridian)	Date of original Norwegian map	Print date of GSGS 4090 paper series	Dimensions (length × width)	Location of extant copies
14D	Kristiania	0°6'E–0°43'W 59°40'–60°N	1900	1940	475 mm × 510 mm	1, 5
15C	Fet	0°5'W–0°54'E 59°41'–59°59'N	1914	1940	475 mm × 510 mm	1, 5
15D	Setskog	0°53'–1°40'E 59°41'–59°59'E	1901	1940	475 mm × 510 mm	1, 5
19B	Hönefoss	0°44'W–0°5'E 59°59'–60°17'N	1917	1940	475 mm × 510 mm	1, 5
19D	Gran	0°45'W–0°4'E 60°17'–60°35'N	1920	1940	475 mm × 510 mm	1, 5
20A	Nannestad	0°5'–0°53'W 59°59'–60°18'N	1912	1940	475 mm × 510 mm	1, 5
20B	Kongsvinger	0°53'–1°42'E 60°–60°18'N	1913	1940	475 mm × 510 mm	1, 5
20C	Eidssvoll	0°4'–0°53'E 60°18'–60°36'N	1910	1940	475 mm × 510 mm	1, 5
20D	Söndre Solör	0°53'–1°42'E 60°18'–60°36'N	1934	1940	475 mm × 510 mm	1, 5
25B	Gjövik	0°46'W–0°3'E 60°35'–60°54'N	1919	1940	475 mm × 510 mm	1, 5
25D	Lillehammer	0°47'W–0°3'E 60°54'–61°12'N	1923	1940	475 mm × 510 mm	1, 5
26A	Hamar	0°17'–0°53'E 60°36'–60°54'N	1912	1940	475 mm × 510 mm	1, 5
26B	Nordre Solör	0°53'–1°42'E 60°35'–60°54'N	1939	1942	425 mm × 577 mm	1, 5
26C	Aamot	0°3'–0°52'E 60°54'–61°12'N	1888	1940	475 mm × 510 mm	1, 5
26D	Söndre Osen	0°17'–0°53'E 60°54'–61°12'N	1933	1942	425 mm × 577 mm	1, 5

Sheet number	Sheet title	Geographical values (based on Oslo meridian)	Date of original Norwegian map	Print date of GSGS 4090 paper series	Dimensions (length × width)	Location of extant copies
31B	Gausdal	0°48'W–0°2'E 61°12'–61°30'N	1904	1940	475 mm × 510 mm	1, 5
E32	Hemsedal	2°30'–1°30'W 60°40'–61°N	1925	1940	490 mm × 335 mm	1, 5
E33 East	Tunhovd	1°30'–2°W 60°20'–60°40'N	1923	1940	490 mm × 335 mm	1, 5
E33 West	Dagali	2°–2°30'W 60°20'–60°40'N	1923	1940	490 mm × 335 mm	1, 5
E34 East	Nore	1°30'–2°W 60°–60°20'N	1936	1940	490 mm × 335 mm	1, 5
E34 West	Maar	2°–2°30'W 60°–60°20'N	1936	1940	490 mm × 335 mm	1, 5
E35 East	Tinnsjö	1°30'–2°W 59°40'–60°N	1931	1940	490 mm × 335 mm	1, 5
E35 West	Rjukan	2°30'–2°W 59°40'–6°N	1930	1940	490 mm × 335 mm	1, 5
F31 East	Synnfjell	0°30'–1°W 61°–61°20'N	1935	1940	490 mm × 335 mm	1, 5
F31 West	Nordre Etnedal	1°–1°30'W 61°–61°20'N	1932	1940	490 mm × 335 mm	1, 5
F32 East	Nordre Land	0°30'–1°W 60°40'–61°N	1930	1940	490 mm × 335 mm	1, 5
F32 West	Aurdal	1°–1°30'W 60°40'–61°N	1930	1940	490 mm × 335 mm	1, 5
F33 East	Sperillen	0°30–1°W 60°20'–60°40'N	1918	1940	490 mm × 335 mm	1, 5
F33 West	Flaa	1°–1°30'W 60°20'–60°40'N	1918	1940	490 mm × 335 mm	1, 5
F34 East	Tyristrand	0°30'–1°W 60°–60°20'N	1919	1940	490 mm × 335 mm	1, 5
F34 West	Sigdal	1°–1°30'W 60°–60°20'N	1935	1940	490 mm × 335 mm	1, 5
F35 East	Eiker	0°30'–1°W 59°40'–60°N	1919	1940	490 mm × 335 mm	1, 5
F35 West	Flesburg	1°–1°30'W 59°40'–60°N	1935	1940	490 mm × 335 mm	1, 5

APPENDIX 5
[Series 43]

Like many other escape and evasion series produced by MI9, this series carried no title or individual sheet names. While based on the existing maps of the International Map of the World (IMW) of the European area, the ten basic sheets were all produced by panelling together sections of existing IMW sheets to produce irregular size sheets, all of which were printed on man-made fibre. The sheets are all at 1:1,000,000 scale, with three of them carrying larger scale insets of border areas. The sheets are all prefixed 43, followed by an alphabet letter. This Appendix provides details of the sheets, scale, geographical coverage, dimensions, the combinations produced, the size of the print runs, the print dates and the production suspension dates. Six sheets in this series were printed in substantial numbers, probably for operational use as well as for escape and evasion purposes. The final column indicates the location of the extant copies of the sheets which have been found.

Sheet number	Scale	Geographical coverage	Dimensions (length × width)
43A	1:1,000,000	NW France, W and C Belgium, part of Holland	734 mm × 736 mm
	insets 1:500,000	2 insets of Pyrenees	
43B	1:1,000,000 inset 1:300,000	SW France, N Spain Inset of German–Swiss frontier	734 mm × 736 mm
43C	1:1,000,000	Holland, C and E Belgium, NE France, W and C Germany	732 mm × 739 mm
43D	1:1,000,000	SE France, SW Germany, NW Italy, Switzerland	732 mm × 739 mm
	inset 1:250,000	Inset of Belgium–German frontier area	
43E	1:1,000,000	N Germany, Bohemia, Moravia, Slovakia, Poland, N Hungary	779 mm × 875 mm
43F	1:1,000,000	W and C Croatia, W Montenegro, S Slovakia, S Germany, N and C Italy, E Switzerland	779 mm × 875 mm
43G	1:1,000,000	S Slovakia, S Poland, SE Hungary, Romania, Serbia, NC and E Bulgaria, E Croatia, E Montenegro, N Albania	761 mm × 917 mm
43H	1:1,000,000	Greece, Albania, S Bulgaria, parts of Turkey Inset of Crete	761 mm × 917 mm
43K East	1:1,000,000	SW France, NE Spain	522 mm × 605 mm
43K West	1:1,000,000	N Portugal, NW Spain	522 mm × 605 mm

Combinations produced	Print runs	Print dates	Production suspended	Location of extant copies
43A/B	350,000	1943, 1944	25.4.45	1, 5
43C/D	300,000	1943, 1944	11.5.45	1, 5, 6
43C/E	60,000	1944. 1945	25.4.45	5
43D/F	10,000			1
43E	4,500			
43E/F	35,000	1943, 1944	11.5.45	1, 3, 5
43F/G	15,000		11.5.45	1, 5
43G/H	35,000	1943, 1944	25.4.45	1, 3, 5
43K East/West	350,000	1943, 1944	11.5.45	1

Total number of sheets produced = 10
Total number of copies printed = 1,159,500

APPENDIX 6
[Series 44]

Identical in specification to [Series 43] were the eighteen sheets, produced in nine set combinations, of the Far East area. Again they were based on the existing IMW sheets, were small-scale and were all produced on man-made fibre. They appear also to have been produced for both escape and evasion and operational purposes. The sheets are all prefixed 44, followed by an alphabet letter. This Appendix provides details of the sheets, scale, geographical coverage, dimensions, the nine set combinations, the size of the print runs, print dates and production suspension dates. The final column indicates the location of the extant copies of the sheets which have been found. While a detailed investigation into the actual use of this particular group of maps lies beyond the scope of this study, future research on the practical uses made of them by the South East Asian Command in its campaigns against Japan might prove to be a worthwhile undertaking.

Sheet number	Scale	Geographical coverage	Dimensions (length × width)
44A	1:1,000,000	NW Burma and part of India	975 mm × 623 mm
44B	1:1,000,000	NE Burma and N Siam and parts of French Indo-China and China	975 mm × 623 mm
44C	1:1,000,000	S Burma, W and C Siam and part of French Indo-China	619 mm × 970 mm
44D	1:1,000,000	S Burma and S Siam	970 mm × 619 mm
44E	1:1,000,000	N Sumatra and part of Siam	980 mm × 633 mm

Sheet number	Scale	Geographical coverage	Dimensions (length × width)
44F	1:1,000,000	Part of Siam, Malaya and E Sumatra	980 mm × 633 mm
44G	1:1,000,000	S Sumatra and NW Java	627 mm × 967 mm
44H	1:1,000,000	3 insets: 1) SW Borneo 2) E Java 3) W Java	627 mm × 967 mm
44J	1:1,000,000	C French Indo-China and E Siam	634 mm × 987 mm
44K	1:1,000,000	S French Indo-China, part of Thailand Inset in SE corner to complete coverage of land area	634 mm × 987 mm
44L	1:1,000,000	Parts of S China and NE French Indo-China	971 mm × 621 mm
44M	1:1,000,000	Part of SW China and N French Indo-China	971 mm × 621 mm
44N	1:1,000,000	Part of mainland China and Hong Kong and Taiwan	851 mm × 1,081 mm
44O	1:1,000,000	Part of C mainland China	851 mm × 1,081 mm
44R	1:1,000,000	Part of mainland China	839 mm × 1,039 mm
44S	1:1,000,000	Korea	1,039 mm × 839 mm
44T		these two sheets were apparently in production in 1945 but were cancelled at proof stage and not printed	
44U			
44V	1:1,000,000	S Japan Insets A and B to complete coverage of land areas	853 mm × 1,028 mm
44W	1:1,000,000	N Japan Inset to complete coverage of land area	853 mm × 1,028 mm

Combinations produced	Print runs	Print dates	Production suspended	Location of extant copies
44A/44B	60,000	21.4.44	21.8.45	1, 5
44C/44D	10,000	13.3.44		1, 5, 6
	20,000	?.6.44	21.8.45	
44E/44F	15,000	17.3.44	21.8.45	1, 5, 6
44G/44H	20,000	11.7.44	31.8.45	1, 5
44J/44K	20,000	7.6.44	21.8.45	1, 5
44L/44M	10,000	7.6.44		1, 5
44N/44O	20,000	25.4.45		1, 5, 6

Combinations produced	Print runs	Print dates	Production suspended	Location of extant copies
44R/44S	5,068	22.9.45		1, 5, 6
44V/44W	5,000	4.10.45		1, 5

Total number of sheets produced = 18 (in 9 set combinations)
Total number of copies produced = 185,000+

APPENDIX 7
[Series FGS]

This series comprises five sheets, produced singly and in various combinations, and again based on the existing IMW series at small-scale, either 1:1,000,000 or 1:1,250,000. All were produced on man-made fibre. The sheets are all prefixed FGS, followed by an alphabet letter. The significance of FGS is not known. This Appendix provides details of the sheets, scale, geographical coverage, the nine combinations produced, the size of the print runs, which vary considerably from as small as 250 copies to as many as 15,400, and the print dates. The final column indicates the location of the extant copies of the sheets which have been found.

Sheet number	Scale	Geographical coverage
FGS A	1:1,000,000	Inset 1: S Norway and adjacent Sweden
	1:1,000,000	Inset 2: C Norway and adjacent Sweden
FGS B	1:1,000,000	S Norway and adjacent Sweden
FGS C	1:1,250,000	N Sweden, N Finland and adjacent USSR
FGS D	1:1,250,000	C and N Norway and adjacent Sweden
FGS E	1:1,000,000	Inset 1: N Norway and adjacent Sweden, Finland, USSR
	1:1,000,000	Inset 2: S Sweden, Denmark, N seaboard of Germany

Combinations produced	Print runs		Print dates	Location of extant copies
FGS A	250	fabric	16.11.42	
FGS A/B	250	paper	16.11.42	
	6,500	fabric	21.12.42	4, 5
	10,000	fabric	24.2.43	
	10,000	unspecified	29.5.43	
	6,000	unspecified	29.7.43	

Combinations produced	Print runs		Print dates	Location of extant copies
FGS A/E	15,400	fabric		1, 3, 5
	10,000	unspecified	21.4.44	
FGS B	250	paper	16.11.42	
FGS C	250	paper	1.12.42	
FGS D	250	paper	1.12.42	
FGS C/D	6,750	fabric	23.12.42	4, 5
	250	fabric	1.12.42	
	10,000	fabric	24.2.43	
	10,000	unspecified	29.5.43	
FGS E	1,000	unspecified	13.4.43	4
FGS E/9CA*	15,000	rayon fabric	25.4.43	1, 5
	3,000		?.4.43	

* refer to Appendix 8 for details of sheet 9CA

Total number of sheets produced = 5
Total number of combinations identified = 9
Total number of copies produced = 105,150

APPENDIX 8
Miscellaneous maps

This Appendix comprises a group of sixteen miscellaneous maps which have been identified, either through mention in the records or by discovering extant copies. In some cases, it is clear that these were produced, initially at least, for inclusion in MI9's *Bulletin* as well as for further operational purposes. The following Appendix lists the sheets and shows, at Part 1, the sometimes limited details which are known of them in terms of sheet number or title, scale, geographical coverage, dates, dimensions and other identifying detail. Part 2 shows production details, specifically whether they were produced singly or in combination, the print runs and print dates. The final column indicates the location of the extant copies of the sheets which have been found.

Part 1: geographical description of the maps

Sheet number/ title	Scale	Geographical coverage	Date	Dimensions (length × width)	Notes
AL1		no known details			
AL2		no known details			
AL3					Possibly titled Liège

Sheet number/ title	Scale	Geographical coverage	Date	Dimensions (length × width)	Notes
CD/A		no known details			
CD/B		no known details			
CD/C		no known details			
CD/D		no known details			
9CA	1:2,350,000	France		789 mm × 528 mm	3 colours marked ADI (Maps) Air Ministry No.7329A
9J3	0.5 in. = 10 miles [1:1,267,200]	Title: Northern Italy Extends N to Switzerland and E to Yugoslavia	Magnetic variation at 1941	505 mm × 590 mm	Monochrome map marked ADI (Maps) Air Ministry No.7331
9J4	1:2,250,000	Title: Southern Italy Extends to cover Sicily plus inset of Sardinia at same scale	Magnetic variation at 1941	505 mm × 590 mm	Monochrome map marked ADI (Maps) Air Ministry No.7332
9S	0.5 in. = 10 miles [1:1,267,200]	Title: Greece	Magnetic variation at 1941	610 mm × 480 mm	Monochrome map marked ADI (Maps) Air Ministry No.7334
9T	1:2,000,000	Title: Bulgaria–Roumania		610 mm × 480 mm	Monochrome map marked ADI (Maps) Air Ministry No.7317
9U	1:3,000,000	Covers Holland, Belgium, Luxembourg, Switzerland, Austria, Poland, Hungary, E to Russian border and N to Copenhagen	Legend shows boundaries at 1939 and 1941	380 mm × 480 mm	Monochrome map marked ADI (Maps) Air Ministry No.7330
9V	1 in. = 47 miles [1:3,000,000]	Title: Stalag locations in northern France		310 mm × 305 mm	Monochrome map marked MI9B and SECRET
Norway	1 in. = 32 miles [1:2,027,520]	Legend indicates map shows in red east and west zones garrisoned by German troops		410 mm × 310 mm	Marked Appendix A & AD Maps AM No. 414/2A
Zones of France	1:2,000,000	France			Monochrome map marked I.S.9 (WEA)

Part 2: *production details and location of extant copies*

Combinations produced	Print runs	Print dates	Location of extant copies
AL1	100 tissue	12.4.42	
	100	?.4.43	
	200 MLS	15.12.43	
AL2	100 tissue	12.4.42	
	100	?.4.43	
	200 MLS	15.12.43	

Combinations produced	Print runs	Print dates	Location of extant copies
AL3	100 tissue	12.4.42	
	100	?.4.43	
	200 MLS	15.12.43	
CD/A	not known	22.7.43	
CD/B	not known	22.7.43	
CD/C	not known	22.7.43	
CD/D	not known	22.7.43	
9CA/9U	5,000 fabric	9.4.43	
9J3	100 tissue	1.2.43	
9J4	100 tissue	1.2.43	
9J3/9J4	silk		1
9S/9T	10,000 fabric	12.4.43	
	2,000 fabric	21.5.43	
	silk		1
9U	silk		1, 6
9V	silk		1
Norway	silk		1
Zones of France	silk		1

Total number of sheets identified = 16
Total number of copies produced = 20,050

APPENDIX 9

Maps produced for the Bulletin

This Appendix comprises a group of eight maps which were initially produced for inclusion in the *Bulletin* and, as such, were printed on standard printing paper. It is believed that the maps were subsequently reproduced on silk or tissue for use in escape and evasion. However, there are known to be some key differences, not least in the sheet numbering. At least one of the maps in the *Bulletin* carries the same number as a very different map produced for escape and evasion. Additionally, there are maps in the *Bulletin* for which extant copies of the escape and evasion versions have not been identified, although there is ample evidence to confirm that they were produced. It is, therefore useful to cross compare entries in this Appendix with the same sheet numbers in Appendix 1 or the same sheet titles in Appendix 8.

Sheet number	Sheet title	Scale	Geographical coverage	Dimensions (length × width)	Detail	Notes	*Bulletin* details
A1	Schaffhausen Salient (West)	1:100,000	German–Swiss border area in region of Schleitheim	147 mm × 175 mm	Red frontier and pylons, black railways, blue water, green woods, hachures relief	Identical in all respects to A1 at Appendix 1	Map No. 4 Chapter 15 Germany
A2	Schaffhausen Salient (East)	1:100,000	German–Swiss border area in region of Ramsen and Engen	200 mm × 147 mm	Red frontier and pylons, black railways, blue water, green woods, hachures relief	Not same map as A2 at Appendix 1	Map No. 5 Chapter 15 Germany Ground photo also shown.
A3	Danzig Docks	(c.1:20,000)	Port of Danzig at mouth of River Vistula	172 mm × 380 mm	Black detail, blue water	Similar to A4 at Appendix 1 but less detail and smaller scale	Plan No. 1 Chapter 15 Germany
A10	Gdynia Docks	n/s	Docks at Gdynia	175 mm × 205 mm	Monochrome Black rail, roads, paths, tree symbols	No similar map identified as E & E version	Plan No. 2 Chapter 15 Germany
A11	Stettin	1:25,000	Stettin and adjacent area with insets of Freihafen and location map of area	395 mm × 280 mm	Extract of native large-scale map with red numbers added to key	No similar map identified as E & E version	Plan No. 3 Chapter 15 Germany
	Lübeck	n/s	Lübeck and adjacent area	217 mm × 270 mm	Extract of native large-scale map	No similar map identified as E & E version	Plan No. 4 Chapter 15 Germany
	Norway Military Zones	(c.1:2,000,000)	Norway S of Trondheim to Swedish border showing E and W Military Zones	330 mm × 250 mm	Black and red detail Dated March 1943	Similar to Norway at Appendix 8	Marked SECRET Appendix C Map B
	Northern Norway	1:1,600,000	Norway N of Trondheim in 2 sections, central and north	430 mm × 465 mm	Black and red detail Dated March 1943	No similar map identified as E & E version	Marked SECRET Appendix C Map C

Total number of maps = 8

Location of extant copies of all eight maps = 2

APPENDIX 10
Decoding a hidden message

The text of the letter dated 22/12/42 (see page 140), which John Pryor wrote home, is reproduced below and the exercise to decode the letter is marked up at page 243 where significant words and letters are highlighted in red. An explanation of how the letter was decoded then follows.

John Pryor's coded letter

22/12/42

My dear Mummy and Daddy, The camps appearance is looking quite smart now as the main road and paths inside the wire have been lined with small trees. Also our keen gardeners have dug flowerbeds in front of each occupied barrack. We were however forced to bring better soil in as most of our camp ground boasts only of sand in which nothing much will grow. Next spring when the new plants are on the way it should look quite respectable. We have just been working hard opening up our Xmas food parcels for this festive week, inside they contain several Xmas luxuries. Some have probably been on view to next-of-kin at the Red cross centres. The parcels are certainly up to standard. Two or three days back a letter came from the Odell's; Alasdair apparently, has joined up and possesses Robert's great liking for high speed travel on the roads. I have not played bridge recently, but hope of a rubber soon. The new five-suit game sounds the most complicated affair. My small model is really well underway and shows quite definite signs by now of resembling a real whaler. The contents of rubbish dumps etc. are really just the thing for getting the odd little bits of wood and tin for it! As regards the clothing parcel suggestions. There is nothing I need really. But you ^already^ understand the few odd consumable things that one needs. I shall really be quite content even if they are underweight. Recently we had a large number of books, but mine have not arrived yet.

Heaps & heaps of love
your loving son
John

John Pryor's coded letter with hidden message highlighted

22/12/42

My dear Mummy and Daddy, The camps **(3 x 5 grid)** appearance is looking quite smart now as the main **road (5)** and paths inside **the (4 – start alphabetical code)** wire have been lined with small trees. **A**lso **o**ur **k**een **g**ardeners **h**ave **d**ug flowerbeds **i**n **f**ront **o**f **e**ach **o**ccupied **b**arrack. **W**e **w**ere **h**owever **f**orced **t**o **b**ring **b**etter **s**oil **i**n **a**s **m**ost **o**f **o**ur **c**amp **g**round **b**oasts **o**nly **o**f **s**and **i**n which nothing much will grow. Next spring when the **new (5)** plants are on **the (4 – start alphabet code)** way it should look quite respectable. **W**e **h**ave **j**ust **b**een **w**orking **h**ard **o**pening **u**p **o**ur **X**mas **f**ood **p**arcels **f**or **t**his **f**estive **w**eek, **i**nside **t**hey **c**ontain **s**everal **X**mas luxuries. Some have probably been **on (5)** view to next-of-kin **at (4)** the Red cross centres. **The (5 – start alphabet code)** parcels are certainly up to standard. **T**wo **o**r **t**hree **d**ays **b**ack **a l**etter **c**ame **f**rom **t**he **O**dell's; **A**lasdair **a**pparently **h**as **j**oined **u**p **a**nd **p**ossesses **R**obert's **g**reat **l**iking for high speed travel on the roads. I have not played **bridge (5)** recently, but hope **of (4)** a rubber soon. The **new (5)** five-suit game sounds **the (4 – start alphabet code)** most complicated affair. **M**y **s**mall **m**odel **i**s **r**eally **w**ell **u**nderway **a**nd **s**hows **q**uite **d**efinite **s**igns **b**y **n**ow **o**f **r**esembling **a r**eal whaler. The contents of rubbish **dumps (5)** etc. are really **just (4)** the thing for getting **the (5 – start alphabet code)** odd little bits of wood and tin for it! **A**s regards the **c**lothing **p**arcel **s**uggestions. **T**here **i**s **n**othing **I n**eed really. **B**ut **y**ou ^**a**lready^ **u**nderstand **t**he **f**ew **o**dd **c**onsumable **t**hings that **o**ne **n**eeds. **I** shall **r**eally be quite content even if they are underweight. Recently we had a **large (5)** number of books, **but (4 – indicates end of message)** mine have not arrived yet.

Heaps & heaps of love
your loving son
<u>John</u>

The first two words after the salutation are 'The camps': a 3 x 5 grid is, therefore, constructed. Moving to the second line of the letter and using Pryor's numerical code of 5 and 4, the fifth word is 'road', so this goes into the top left box of the grid and will, therefore, be the final word of the message.

road		

The fourth word after this is 'the', which indicates that the alphabet code starts at this point. This means that, starting with the next sentence, the first letter of each consecutive word is written down in groups of three. Each letter is identified on the alphabet listing shown above right and the number of the column (1, 2 or 3) in which it occurs is noted.

The decoder, therefore, moves to the next sentence which starts 'Also our keen' and proceeds to list the first letter of every word in groups of three. Each group of three letters signifies a letter in Pryor's alphabet table, S, as shown above right.

S 111	T 211	U 311
V 112	W 212	X 312
Y 113	Z 213	. 313
A 121	B 221	C 321
D 122	E 222	F 322
G 123	H 223	I 323
J 131	K 231	L 331
M 132	N 232	O 332
P 133	Q 233	R 333

Transposing each group of three letters into the alphabet letter from this table, the following word emerges:

A O K = 132 = M
G H D = 121 = A
F I F = 333 = R
O E O = 323 = I
B W W = 222 = E
H F T = 232 = N
B B S = 221 = B
I A M = 311 = U
O O C = 333 = R
G B O = 123 = G
O S I = 313 = .

The word spells 'marienburg' and becomes the second word on the grid.

At the point where the full stop occurs, the decoder reverts to the 5 4 sequence at the start of the next sentence in the letter but maintaining the correct rhythm. Having finished on the fourth word at the previous stage, this time the fifth word is counted. Starting counting at the beginning of the next sentence, this gives the word 'new', which becomes the third word on the grid.

The following fourth word is 'the' signalling that the alphabet code starts again at the beginning of the next sentence. Taking the first letter of each word and setting them out in groups of three produces the following:

W H J = 221 = B
B W H = 222 = E
O U O = 333 = R
X F P = 331 = L
F T F = 323 = I
W I T = 232 = N
C S X = 313 = .

The fourth word on the grid is 'berlin'.

road	marienburg	new
berlin		

Moving to the start of the next complete sentence in the letter and picking up the 5 4 rhythm, the fifth word is 'on' which becomes the fifth word on the grid.

The following fourth word is 'at', which becomes the sixth word on the grid, noting that the hyphenated 'next-of-kin' counts only as one word.

The following fifth word in the letter is 'the', which indicates that the alphabet code starts again at the beginning of the next sentence.

Taking the first letter of each word and setting them out in groups of three produces the following:

T O T = 232 = N
D B A = 121 = A
L C F = 333 = R
T O A = 231 = K
A H J = 121 = A
U A P = 311 = U
R G L = 313 = .

The seventh word on the grid is 'narkau'.

Starting with the next sentence, the fifth word is 'bridge' which becomes the eighth word. It should be noted that Pryor had lost the 5 4 rhythm count as it should arguably have been the fourth and not the fifth word. These occasional lapses were entirely understandable and the decoders simply tried alternatives (as indeed did the author) when the message appeared to make no sense or started to lose sense.

The following fourth word is 'of', which becomes the ninth word.

The following fifth word is 'new', which becomes the tenth word.

Noting the hyphenated word (five-suit) which counts as one word, the next fourth word is 'the' which again signals the start of the alphabet code at the beginning of the next sentence.

Taking the first letter of each word and setting them out in groups of three produces the following:

APPENDICES 245

M S M = 111 = S
I R W = 332 = O
U A S = 311 = U
Q D S = 211 = T
B N O = 223 = H
R A R = 313 = .

The eleventh word on the grid is 'south'.

road	marienburg	new
berlin	on	at
narkau	bridge	of
new	south	

Moving to the start of the next complete sentence in the letter and keeping the 5 4 rhythm, the fifth word is 'dumps' which becomes the twelfth word on the grid.

The following fourth word is 'just' (noting the inclusion of the otherwise unnecessary 'etc.' to ensure the count identified the correct word), which becomes the thirteenth word on the grid.

The following fifth word is 'the', which indicates the start of the alphabet code at the beginning of the next sentence. Taking the first letter of each word (noting the insertion of 'already' to ensure the correct sequence of letters) and setting them out in groups of three produces the following:

A R T = 132 = M
C P S = 311 = U
T I N = 232 = N
I N R = 323 = I
B Y A = 211 = T

U T F = 323 = I
O C T = 332 = O
T O N = 232 = N
I S R = 313 = .

The fourteenth word on the grid is 'munition'.

Moving to the start of the next complete sentence in the letter, the fifth word is 'large' which becomes the fifteenth and final word on the grid. Arguably it should have been the fourth word 'a' at this point to ensure retention of the 5 4 rhythm: either 'a' or 'large' would make sense: 'large' makes more sense since he used the plural form 'dumps' and the letter writers did not usually include unnecessary words in their hidden messages such as an article before a noun. To reinforce the fact that this was the end of the message, the fourth word after this is 'but'. Placing the words in the correct numerical order 1 to 15 on the grid gives the result shown:

road	marienburg	new
berlin	on	at
narkau	bridge	of
new	south	dumps
just	munition	large

Starting in the bottom right corner and reading across and diagonally in sequence, the message reads:

LARGE MUNITION DUMPS JUST SOUTH OF NEW BRIDGE AT NARKAU ON NEW BERLIN MARIENBURG ROAD

BIBLIOGRAPHY

PRIMARY SOURCES

A recurring aspect of the research which underpins this study is the continuing search for files in The National Archives. Foot and Langley were afforded special access in the late 1970s to produce their seminal work, *MI9: Escape and Evasion, 1939–1945*. Foot anticipated, in the author's discussion with him in 2012, that many of the previously closed files would by now be open. Certainly many more than he was able to access are now available. However, identifying the files can often be testing since they are not all amongst the War Office files, which was where Foot anticipated they would be, and some carry titles and descriptions which do not readily indicate that they relate to the work of MI9. Such shortcomings in cataloguing may well have resulted from the dispersal of files at the time MI9 was being wound up at the end of the war.

Even now, the researcher is still faced with closed files. This author's experience of just one request under the Freedom of Information (FOI) legislation for the opening of a file led to a wait of over three months because the request reportedly raised 'complex public interest considerations'. Government departments generally respond to FOI requests within four weeks (twenty working days). This particular FOI request was made on 8 May 2013 and the file was only finally opened on 15 August 2013, a period of over three months. Moreover, significant segments of the released file had been redacted and the closure dates of some of the redactions had been further extended to 2030. The imagination can run amok, thinking about what possible secrets need to be withheld for eighty-five years after the end of World War II.

British Library
Waddington Company Archive (4 volumes), Maps C.49e.55.
D.Survey World War II print record, Maps UG-9-152-H.

Defence Geographic Centre
File Svy 2/6330. Fabric Maps Pt I 10-8-44 to 30-9-46.

Macclesfield Silk Museum
Brocklehurst Whiston Amalgamated Board Minute Books 1935–1947.

National Library of Scotland
Bartholomew Archive, Accession 10222.

Royal Air Force Museum
Documents B3227 and DB319.

THE NATIONAL ARCHIVES
Air Ministry files
AIR 14/353–61 (9 files), RAF personnel taken prisoner: aids to escape, conduct, etc.
AIR 14/461–4 (4 files), Prisoners of war instructions and methods of escape.

AIR 20/2328, Escape aids policy for prisoners of war (file missing).

AIR 20/9165, Allied prisoners of war: escape aids policy.

AIR 20/6805, RAF personnel: evasion and escape equipment.

AIR 40/2645, Stalag Luft III (Sagan): camp history.

AIR 40/2457, MI9: escape and evasion: post war policy 1944–1946.

AIR 51/260, Intelligence Section escape aids.

AIR 76/247/127, Officers service records of RAF, 1 Jan 1918–31 Dec 1919.

Cabinet Office files

CAB 79/15/30, MI9 War establishment.

Special Operations Executive files

HS 9/771/4, Personnel files 1 Jan 39–31 Dec 46.

Air Ministry and Ministry of Defence papers accumulated by the Air Historical Branch

IIIL 50/1/7 (A), Prisoners of war allied escape aids policy.

Maps and plans

MF1/2 , Maps and plans of NW Europe extracted from files for flat storage.

Domestic records of the Public Record Office

PRO 219/1448, Advice to prisoners of war in 1944.

Treasury Solicitor files

TS 28/580, Advice of litigation arising from publication of material detailing the invention and manufacture of gadgets 1950–1956.

TS 28/581, Advice of litigation arising from publication of material detailing the invention and manufacture of gadgets 1 Jan 1951–31 Dec 1960 .

War Office files

WO 78/5814, Second World war escapers' maps on fabric.

WO 165/39, MI9 war diaries.

WO 169/24879, Records of MI9 organization and activity.

WO 169/29748, Records of MI9 organization and activities 1941–46.

WO 201/1417, Prisoner of war escapes: devices.

WO 208/3242, Historical record MI9 1 Jan 39–31 Dec 45.

WO 208/3243, Historical record MI9 1 Mar 42–28 Feb 46.

WO 208/3244, Reports by RAF prisoners of war in Germany.

WO 208/3245, Reports by RAF prisoners of war in Germany.

WO 208/3246, History of IS9: Western Europe.

WO 208/3247, British prisoners of war in Germany Oct–Dec 1939.

WO 208/3252, Prisoners of war Section E Group Bulletins 1944–45.

WO 208/3255, Disbandment of IS9 May–June 1945.

WO 208/3267, Combined operations: particulars of purses March 42–Sept 43.

WO 208/3268, MI9 Bulletins.

WO 208/3270, Marlag und Milag Nord: camp history.

WO 208/328, Stalag XXA Thorn: camp history.

WO 208/3288, Oflag IVC Colditz 1940–45: camp history.

WO 208/3297, The Escapers' Story: a compilation of various escape reports 1940–41.

WO 208/3298–327 (30 files), Escape and evasion reports.

WO 208/3328–40 (13 files), Liberation reports.

WO 208/3431, Lectures to units by escaped prisoners of war June–Aug 1942.

WO 208/3445, MI9 lectures on conduct if cut off from unit or captured.

WO 208/3450, MI9 Establishment 1939–42.

WO 208/3501, Marlag and Milag Nord escaper and evader coded letter traffic.

WO 208/3502, Oflag VIIC coded letter traffic.

WO 208/3503, Oflag XIIB coded letter traffic Jan 43–Apr 45.

WO 208/3512, Distribution of MI9 reports May–Nov 1940.

WO 208/3554, MI9 special equipment: requirements, approval and supply.

WO 208/3566, English/German coded dictionary.

WO 208/3569, IS No 9 West Europe area: history.

WO 208/3572, MI9 re-organization.

WO 208/3574, Inter-service escape and evasion training: A19 lecture papers.

WO 344/260/2, Liberated prisoners of war interrogation questionnaires.

PRIVATE PAPERS

Imperial War Museum

Typescript of the original draft for Foot and Langley *MI9: Escape and Evasion, 1939–1945,* includes the unpublished footnotes, Miscellaneous Document 2744

Parliamentary Archives

Airey Neave's personal papers, accession number 1907, deposited 1979. General files AN 598–627, Business files AN 628–37, Literary files AN 638–71, War and post-war files AN 672–707.

Second World War Experience Centre

John Pryor's memoirs, Reference SWWEC RN/Pryor J.

PUBLISHED SOURCES

Exploring the published literature on MI9's escape and evasion mapping programme has proved to be a considerable challenge as there is a very real paucity of published sources on the subject. While there is a vast array of published sources relating to the activity of escape and evasion, very little of it mentions maps and, even sources that do, make only passing mention of the subject. Some of the primary published sources proved to be of crucial importance. The key one among these is undoubtedly that written by Christopher Clayton Hutton, *Official Secret*, published in 1960, and it is discussed fully in Chapter 9.

In 1969 Airey Neave published his second book, *Saturday at MI9,* the title of which reflected his own codename (Saturday) in MI9. Neave acknowledged in the preface to his book that it was not possible at that stage to write an official history of MI9, that he was writing rather about his personal experiences working in MI9 and that his was the first to be written 'from the inside', a somewhat curious statement since Hutton's own book had been published some nine years earlier. However, there is no mention in Neave's book of Hutton or of any aspect of the escape aids programme on which he worked. Neave concentrated rather on the work done in support of the escape lines and their organization. While the book is of general interest and describes, albeit in a limited way, some of the educational and training aspects of MI9's work, it is largely not germane to the subject of this study.

The single most comprehensive published secondary source which describes the whole story of the creation and operation of MI9 is undoubtedly Foot and Langley's *magnum opus, MI9 Escape and Evasion 1939–1945,* published in 1979. Maps are certainly mentioned in the book, particularly in relation to Hutton's escape aids, and the authors highlight the extent to which he quickly realized that maps were an indispensable aid to successful escape. However, it is clear that their source for this information was Hutton's own book and not a primary source such as the War Office (WO) files relating to MI9's activity. Indeed, they highlight the fact that many WO files were withheld from them, estimating the number at some 250 in all, and emphasized the likelihood that there may well have been genuine operational reasons for not making such files publicly available until 2010 since there is no point in advertising clandestine methods that might be used to national advantage another time.

During the 1980s and 1990s a number of relevant, illustrated articles appeared in professional publications and the national press. The most notable ones were those produced by the author as a result of her early interest in the subject when working in the Ministry of Defence. The first of these appeared in 1983 and stirred some considerable interest in professional cartographic circles. This was followed a year later by a much more detailed article resulting from a lecture invited by the British Cartographic Society. These have been widely used, quoted and referenced in other articles on the subject. In 1988 John Doll quoted in detail from Bond's published work, as did Debbie Hall writing over ten years later in 1999. Additionally, there were some short articles in the press, often recounting the escape stories of (by now) elderly men, who had been prisoners of war.

There is a plethora of literature which serves to set the general context of escape in World War II and the background against which MI9 was operating. Many tell the tales of heroic exploits and derring-do but sadly do not contribute to the story of the escape maps. Not even the engagingly titled *Silk and Barbed Wire* proved to be directly relevant since it is a collection of personal reminiscences of captured members of Bomber Command, who had fallen into enemy hands after parachuting from doomed aircraft: the 'silk' of the title related rather to their parachutes and not escape maps, which received no mention at all. There is, however, one book which opened up an intriguing aspect of MI9's mapping programme and that is Julius Green's fascinating insight into the codes operating between the potential escapers in the camps and

MI9 in London, *From Colditz in Code*. It was Foot and Langley who highlighted the existence of the coded contact and they used extracts from Green's book to illustrate the practice.

This review would be incomplete without at least a brief mention of a book published in 1921, *The Escaping Club*. A. J. Evans's book describes his personal experience of capture and subsequent escape during World War I, providing both an insight into the changing military philosophy with regard to prisoners of war and also the personal ingenuity and forward planning of professional soldiers, and their families, should they fall into enemy hands. It is a remarkable book in that it provides the background and explanation for the role which Evans subsequently played in World War II as a member of the staff of MI9, and their choice of escape routes, (see Chapters 1 and 6).

Only recently has there appeared any kind of detailed review of the mapping of World War I. Peter Chasseaud's timely and masterly publication *Mapping the First World War* was published in November 2013. In it he plots the course of British military mapping through the passage of the war years and shows that maps went far beyond the simple visual image and, together with the increasing use of aerial photography, allowed for the development of sophisticated artillery target plans. They were designed for a clear military purpose. The only equivalent publication for World War II, to date, was the War Office's own authorized publication *Maps and Survey*. However, published in 1952, it was clearly intended to serve as the official historical record for internal departmental use since it was classified RESTRICTED. It was not declassified until some twenty years later and there are, therefore, very few copies in the public domain. It was one of a series of volumes, compiled by authority of the Army Council, the object of which was to preserve the experience gained during World War II in selected fields of military staff work and administration. It covers just about every aspect of the operational mapping produced by the Geographical Section General Staff (GSGS), with the single notable exception of MI9's escape and evasion mapping programme. It is not surprising to discover this omission since MI9 did not involve the military mapping authority in their work until very late in the proceedings.

PRIMARY PUBLISHED SOURCES

Churchill, Winston S. *London to Ladysmith via Pretoria*. Longmans, Green & Co. London. 1900.

Connell, Charles. *The Hidden Catch*. Elek Books. London. 1955.

Durnford, H.G. *The Tunnellers of Holzminden*. Cambridge University Press. Cambridge. 1920, 2nd edition 1930.

Evans, A. J. *The Escaping Club*. The Bodley Head. London. 1921.

Evans, Michael. 'PoW tells of escape maps printed on secret press.' *The Times*. 23 June 1997.

Evans, P. Radcliffe. 'The Brunswick Map Printers.' *Printing Review*, Special Cartographic Number, Winter 1951–52.

Green, J. M. *From Colditz in Code*. Robert Hale. London. 1971.

Hutton, Christopher Clayton. *Official Secret: the remarkable story of escape aids, their invention, production and the sequel.* Max Parrish. London. 1960.

—. 'Escape secrets of World War II.' *Popular Science.* Vol. 184, No. 1, pp. 69–73, 200.

James, David. *A Prisoner's Progress.* William Blackwood, Edinburgh and London. 1947. Originally published as 'A Prisoner's Progress' in *Blackwood's Magazine,* December 1946 and January/February 1947. Reissued as *Escaper's Progress.* W. W. Norton. New York. 1955.

Langley J. M. *Fight Another Day.* Collins. London. 1974.

Marks, Leo. *Between Silk and Cyanide. A Codemaker's War 1941–45.* HarperCollins. London. 1998.

Neave, Airey. *They Have Their Exits.* Hodder and Stoughton. London. 1953.

—. *Saturday at MI9.* Hodder & Stoughton. London. 1969.

SECONDARY PUBLISHED SOURCES

Books

Bickers, R. Townshend. *Home Run – Great RAF Escapes of the Second World War.* Leo Cooper. London. 1992.

Cantwell, John D. *The Second World War: a Guide to Documents in the Public Record Office.* PRO Handbook No.15, 3rd edn. London. 1993.

Chasseaud, Peter. *Mapping the First World War.* Collins in association with the Imperial War Museum. Glasgow and London. 2013.

Churchill, Sir Winston S. *The Second World War.* Abridged one-volume edition. Cassell. London. 1959.

Clarke, W. *An Introduction to Textile Printing.* Butterworth. London. 1964.

Clough, A. B. *Maps and Survey.* The War Office. London. 1952.

Clutton-Brock, Oliver. *Footprints on the Sands of Time: RAF Bomber Command Prisoners of War in Germany, 1939–1945.* Grub Street. London. 2003.

—. *RAF Evaders: The Complete Story of RAF Escapees and Their Escape Lines, Western Europe, 1940–1945.* Grub Street. London. 2009.

Collins, Louanne. *Macclesfield Silk Museums – a Look at the Collections.* Macclesfield Museums Trust. 2000.

Collins, Louanne, and Stevenson, Moira. *Macclesfield: The Silk Industry.* Chalford. Stroud. 1995.

—. *Silk – 150 years of Macclesfield Textile Designs (1840–1990).* Macclesfield Museums Trust. (no date).

Cooper, Alan W. *Free to Fight Again: RAF Escapes and Evasions 1940–45.* William Kimber. London. 1988.

Deacon, Richard. *A History of the British Secret Service.* Granada. London. 1969.

Dear, Ian. *Escape and Evasion – Prisoner of War Breakouts and the Routes to Safety in World War II.* Arms & Armour Press. London. 1997.

Dear, I. C. B., and Foot, M. R. D. *The Oxford Companion to the Second World War.* Oxford University Press. Oxford. 1995.

Duncan, Michael. *Underground from Posen.* William Kimber. London. 1954.

Foot, M. R. D. *Resistance: European Resistance to Nazism 1940–45.* Eyre Methuen. London. 1976.

—. *SOE: The Special Operations Executive 1940–46.* BBC Books. London. 1984.

Foot M. R. D., and Langley, J. M. *MI9: Escape and Evasion, 1939–1945.* Bodley Head. 1979.

Gaddum, Anthony H. *Gaddums Revisited.* Amadeus Press ('for private circulation only'). Cleckheaton. 2005.

Gardiner, Leslie. *Bartholomew, 150 years.* John Bartholomew & Son Ltd. Edinburgh. 1976.

Harley, J. B., and Woodward, David. *The History of Cartography.* Volume Two. The University of Chicago Press. Chicago. 1994.

Haynes, Alan. *Invisible Power: The Elizabethan Secret Services 1570–1603.* Alan Sutton. Bath. 1992.

Hinsley, F. H., Thomas, E. E., Ransom, C. F. G., and Knight, R. C. *British Intelligence in the Second World War.* Vol. 1. HMSO. London. 1979.

Hinsley, F. H. *British Intelligence in the Second World War: Its Influence on Strategy and Operations.* Vol. 2. HMSO. London. 1981.

Hinsley, F. H., Thomas, E. E., Ransom, C. F. G. and Knight, R. C. *British Intelligence in the Second World War: Its Influence on Strategy and Operations.* Volume 3, Part 1. HMSO. London. 1984.

Hinsley, F. H., Thomas, E. E., Simkins, C. A. G., and Ransom, C. F. G. *British Intelligence in the Second World War: Its Influence on Strategy and Operations.* Vol. 3, Part 2. HMSO. London. 1988.

Hinsley, F. H., and Simkins, C. A. G. *British Intelligence in the Second World War: Security and Counter-Intelligence.* Vol. 4. HMSO. London. 1990.

Hinsley, F. H., and Howard, Michael. *British Intelligence in the Second World War: Strategic Deception.* Vol. 5. HMSO. London. 1990.

Kain, Roger J. P., and Oliver, Richard R. *The Tithe Maps of England and Wales: a cartographic analysis and county-by-county catalogue.* Cambridge University Press. Cambridge. 1995.

Keene, Thomas Edward. Beset by Secrecy and Beleaguered by Rivals: the Special Operations Executive and Military Operations in Western Europe 1940–1942 with special reference to Operation Frankton. Unpublished University of Plymouth PhD thesis. 2011.

Nichol, John, and Rennell, Tony. *The Last Escape. The untold story of Allied prisoners of war in Germany 1944–45.* Viking. London. 2002.

Report of the Committee on a Military Map of the United Kingdom together with the minutes of evidence and appendices. War Office. London. 1892.

Ryan, Cornelius. *A Bridge Too Far.* Hamish Hamilton. London. 1974.

Seymour, W. A. *A History of the Ordnance Survey.* Wm. Dawson & Sons. Folkestone. 1980.

Singh, Simon. *The Code Book. The Secret History of Codes and Code Breaking.* Fourth Estate. London. 2000.

Storey, Joyce. *Textile Printing.* Thames and Hudson. London. 1974.

Teare, D. *Evader.* Air Data Publications. 1996.

Walley, Brian. *Silk and Barbed Wire.* Sage Pages. Warwick, Western Australia. 2000.

Wallis, Helen, and Robinson, Arthur. *Cartographical Innovations.* Map Collector Publications, Tring. 1987.

Watson, Victor. *The Waddingtons Story.* Northern Heritage Publications. Huddersfield. 2008.

West, Nigel. *Secret War.* Hodder and Stoughton. London. 1992.

Yee, Cordell D. K. 'Reinterpreting Traditional Chinese Geographical Maps' in *The History of Cartography,* Vol. 2.2, pp.35–70. Chicago University Press. Chicago. 1994.

Articles

Balchin, W. G. V. 'United Kingdom Geographers in the Second World War: a Report'. *The Geographical Journal,* Vol. 153, No. 2, July 1987, pp.159–80.

Baldwin, R. E. 'Silk escape maps: where are they now?' *Mercator's World,* Jan/Feb 1998, pp.50–51.

Bond, Barbara A. 'Maps printed on silk'. *The Map Collector,* No. 22, March 1983, pp.10–13.

—. 'Silk Maps: the story of MI9's excursion into the world of cartography 1939–1945'. *The Cartographic Journal,* Vol. 21, No. 2, 1984, pp.141–3.

—. 'Escape and evasion maps in WWII and the role played by MI9'. *The Ranger,* Vol. 2, No. 9, Summer 2009.

Collier, Peter. 'The Work of the British Government's Air Survey Committee and its Impact on Mapping in the Second World War'. *The Photogrammetric Record,* Vol. 21, Issue 114, June 2006, pp.100–9.

Doll, John G. *Cloth Maps of World War II.* Western Association of Map Libraries (WAML), Information Bulletin 20(1), November 1988.

Du, Daosheng. 'The Science of Cartography in China'. *The Cartographic Journal,* Vol. 21, No. 2, December 1984, pp.145–7.

Hall, Debbie. *Wall Tiles and Free Parking: Escape and Evasion Maps of World War II.* British Library. London. April 1999.

Heffernan, Michael. 'Geography, Cartography and Military Intelligence: the Royal Geographical Society and the First World War'. *Transactions of the Institute of British Geographers,* New Series, Vol. 21, No. 3, 1996, pp.504–33.

Heffernan, Michael. 'The Politics of the Map in the Early Twentieth Century'. *Cartography and Geographical Information Science,* Vol. 29 No. 3, 2002, pp.208–26.

Hsu, Mei-Ling. 'The Han Maps and Early Chinese Cartography'. *Annals of the Association of American Geographers,* Vol. 68, 1978, pp.45–60.

Jin, Yingchun. 'China's Achievements in Surveying and Mapping Techniques during the Han Dynasty as Reflected by the Silk Maps Unearthed from the Han Tomb at Mawangdui, China'. *ICA Proceedings,* Warsaw. (See also 'Report on the Military map discovered in Han Tomb Three at Mawangdui'. *Wenwu,* Vol. 1, 1976, pp.18–23 and Zhan Libo 'Notes on the Military map from the Han Tombs at Mawangdui'. *Wenwu,* Vol. 1, 1976, pp. 24–7.)

Nicholas, William H. 'The making of military maps'. *The National Geographic Magazine,* Vol. 33, No. 6, June 1943, pp.764–78.

Pryor, Stephen. 'Obituary: John Pryor'. *The Old Oundelian,* Summer 2012, p.126.

Stanley, Albert A. 'Cloth maps and charts'. *The Military Engineer,* Vol. 39, 1947, p.126.

ILLUSTRATION AND IMAGE CREDITS

References to map scales in the book relate to that of the original map as produced and not necessarily the scale of the illustration as shown.

Page 11 Library of Congress Geography and Map Division, Washington, DC, G3920 1864. M4 Copy 2 Vault Cloth: CW 129.74
Pages 12–13 from *The Escaping Club,* A. J. Evans
Page 14 © Popperfoto / Contributor (Getty)
Page 15 © Leonard McCombe / Stringer (Getty)
Page 16 from *Per Ardua Libertas*
Page 18 Parliamentary Archives AN707
Page 20 from *Official Secret,* C. C. Hutton
Page 22 (top) from *The Escaping Club,* A. J. Evans
Page 22 (bottom) from *The Escaping Club,* A. J. Evans
Page 23 from *Fight Another Day,* J. M. Langley
Page 24 from *They Have Their Exits,* Airey Neave
Page 27 by courtesy of the Trustees of the Royal Air Force Museum
Page 28–29 from *The Tunnellers of Holzminden,* H. G. Durnford
Page 32 reproduced with thanks to the Hunan Provincial Museum, Changsha, China
Page 35 reproduced with thanks to the Bartholomew Family
Page 36 courtesy of the Trustees of the National Library of Scotland
Page 37 courtesy of the Trustees of the National Library of Scotland
Page 38 reproduced with thanks to the Hunan Provincial Museum, Changsha, China
Page 39 reproduced with thanks to the Hunan Provincial Museum, Changsha, China
Page 40 Library of Congress Geography and Map Division, Washington, DC, G3920 1864. M4 Copy 2 Vault Cloth: CW 129.74
Page 41 Library of Congress Geography and Map Division, Washington, DC, G3920 1864. M4 Copy 2 Vault Cloth: CW 129.74
Page 42 Illustration provided by and reproduced with the permission of the owner
Page 44 Illustration provided by and reproduced with the permission of the owner
Page 46 Illustrations provided by and reproduced with the permission of the owner
Page 47 Illustration provided by and reproduced with the permission of the owner
Page 48 Illustration provided by and reproduced with the permission of the owner
Page 49 Illustrations provided by and reproduced with the permission of the owner
Page 50 Illustration provided by and reproduced with the permission of the owner
Page 51 Illustration provided by and reproduced with the permission of the owner
Page 53 Illustration provided by and reproduced with the permission of the owner
Page 55 Illustration provided by and reproduced with the permission of the owner
Page 56 Illustrations provided by and reproduced with the permission of the owner
Page 57 Illustrations provided by and reproduced with the permission of the owner
Page 59 Illustration provided by and reproduced with the permission of the owner
Page 61 © UK MOD Crown Copyright, 2015
Page 62 by courtesy of the Trustees of the Royal Air Force Museum
Page 63 by courtesy of the Trustees of the Royal Air Force Museum
Page 64 by courtesy of the Trustees of the Royal Air Force Museum
Page 65 by courtesy of the Trustees of the Royal Air Force Museum
Page 66 (top) Illustration provided by and reproduced with the permission of the owner
Page 66 (bottom) © Keystone / Staff (Getty)
Page 67 Illustration provided by and reproduced with the permission of the owner
Pages 68–69 Illustration provided by and reproduced with the permission of the owner

Page 71 Illustration provided by and reproduced with the permission of the owner
Page 72 Illustration provided by and reproduced with the permission of the owner
Page 74 by courtesy of the Trustees of the Royal Air Force Museum
Page 75 by courtesy of the Trustees of the Royal Air Force Museum
Page 77 © The National Archives
Page 78 © Victor Watson, reproduced with the permission of the owner
Page 80 © Victor Watson, reproduced with the permission of the owner
Pages 84–85 by courtesy of the Trustees of the Royal Air Force Museum
Page 88 by courtesy of the Trustees of the Royal Air Force Museum
Page 89 © Imperial War Museum (Art.IWM ART LD 5134)
Page 90 © Imperial War Museum (EPH 2217)
Page 92 © Woodmansterne Publications, Watford/TopFoto.co.uk
Page 93 from *Per Ardua Libertas*
Page 94 by courtesy of the Trustees of the Royal Air Force Museum
Page 95 redrawn from an original supplied by courtesy of the Trustees of the Royal Air Force Museum
Page 96 (top) by courtesy of the Trustees of the Royal Air Force Museum
Page 96 (bottom) by courtesy of the Trustees of the Royal Air Force Museum
Page 97 © Imperial War Museum (HU 20276)
Page 98 Reproduced by permission © The Cumberland Pencil Company
Page 99 (bottom left) © Woodmansterne Publications, Watford/Topfoto.co.uk
Page 99 (bottom right) by courtesy of the Trustees of the Royal Air Force Museum
Page 100 by courtesy of the Trustees of the Royal Air Force Museum
Page 101 Image courtesy of Plymouth University and with the permission of John Shelmerdine's daughter
Page 102 by courtesy of the Trustees of the Royal Air Force Museum
Page 103 from *MI9 Escape and Evasion 1939-1945*, M. R. D. Foot & J. M. Langley
Page 105 from *Per Ardua Libertas*
Page 106 from *Per Ardua Libertas*
Page 108 © Imperial War Museum (HU 20926)
Pages 110–111 © UK MOD Crown Copyright, 2015
Page 114 Image courtesy of Plymouth University and with the permission of John Pryor's son
Page 117 from *From Colditz in Code*, J. M. Green
Page 119 from *From Colditz in Code*, J. M. Green
Page 121 from *From Colditz in Code*, J. M. Green
Page 122 Image courtesy of Plymouth University and with the permission of John Pryor's son
Page 123 © Imperial War Museum (HU 20930)
Page 124 by courtesy of the Trustees of the Royal Air Force Museum
Page 126 © Fox Photos / Stringer (Getty)
Page 132 © Imperial War Museum (D 20928)
Page 133 © Imperial War Museum (D 20387)
Page 139 Image courtesy of Plymouth University and with the permission of John Pryor's son
Page 140 Image courtesy of Plymouth University and with the permission of John Pryor's son
Page 142 © Hulton Archive/Stringer (Getty)
Page 144 © Time Life Pictures / Contributor (Getty)
Page 146 © British Library Board
Page 148 from *Fight Another Day*, J. M. Langley
Pages 150–151 Illustration provided by and reproduced with the permission of the owner
Page 152 © British Library Board
Page 153 © British Library Board
Pages 154–155 © British Library Board
Page 157 (top) by courtesy of the Trustees of the Royal Air Force Museum

Page 157 (bottom left and bottom right) © The National Archives
Page 158 from *The Escaping Club,* A. J. Evans
Page 161 © Imperial War Museum (HU 20288)
Page 162 from *They Have Their Exits,* Airey Neave
Page 163 produced by the GeoMapping Unit, Plymouth University and reproduced with their kind permission
Page 164 Parliamentary Archives AN707
Page 168 Parliamentary Archives AN670
Page 170 © Keystone / Staff (Getty)
Page 172 © The National Archives
Page 174 Illustration provided by and reproduced with the permission of the owner
Page 175 © The National Archives
Pages 176-177 Illustration provided by and reproduced with the permission of the owner
Page 180 Image courtesy of Plymouth University and with the permission of John Pryor's son
Page 181 © Imperial War Museum (Art.IWM ART LD 5114)
Page 182 © The National Archives
Page 183 © Imperial War Museum (BU 4835)
Page 184 from *A Prisoner's Progress,* David James
Page 185 from *A Prisoner's Progress,* David James
Page 187 © Imperial War Museum (EPH 6369)
Page 188 by courtesy of the Trustees of the Royal Air Force Museum
Page 190 © Imperial War Museum (BU 5985)
Page 192 (bottom left) Illustration provided by, and reproduced with, the permission of the University of Glasgow
Page 192 (bottom right) Illustration provided by and reproduced with the permission of the owner
Page 193 Illustration provided by and reproduced with the permission of the owner
Page 194 Illustration provided by, and reproduced with, the permission of the University of Glasgow
Page 195 produced by the GeoMapping Unit, Plymouth University and reproduced with their kind permission
Page 196 from *Per Ardua Libertas*
Page 199 from *Official Secret,* C. C. Hutton
Page 204 © Leemage / Contributor (Getty)
Page 207 © Hulton Archive / Stringer (Getty)
Page 212 Illustration provided by and reproduced with the permission of the owner

Front Cover - dust jacket (photo) © Hulton Archive/Stringer (Getty), (map extract Berlin - Sheet 43G) provided by and reproduced with the permission of the owner
PLC (map extract Shaffhausen - Sheet 43B)

Whilst every effort has been made to trace the copyright holders, in cases where this has been unsuccessful, or if any have inadvertently been overlooked, the Publishers would gladly receive any information enabling them to be rectified at the first opportunity.

The author and publishers would finally like to thank Christopher Riches for his invaluable editorial help in making this book.

INDEX

1st Searchlight Regiment 24, 160
2nd Battalion Coldstream Guards 23

A. G. Spalding & Bros 89
Africa 34, 52, 53, 59, 71
Air Ministry 34, 54, 201, 209
Alabama 10, 40
Alger 125
Allied Paper Merchants 81
Alston, E. D. 78, 80, 81, 92
Arnhem 46, 60, 63
Asia, southeast 53, 72
Assistant Directorate of Intelligence (Logistics) 75
Atlanta 10, 40
Austria 46, 193
Author's Society 106

Babington, Anthony 116
backgammon 89
Baden-Powell, R. S. S. 33
Bader, Douglas 30
Baghdad 82
Baileys 93
Balchin, W. G. V. 7, 197, 198
Baltic Coast 175
Baltic ports 9, 76, 87, 125, 139, 148, 173–188
Barcelona, 205
Barry, Captain R. F. T. 169
Bartholomew *see* John Bartholomew & Son Ltd
Bartholomew, John 34
Bartholomew, John "Ian" 34, 35, 52, 54

Batavia (Jakarta) 72
BBC 181
Beethoven 97
Beirut 82
Belgian–German frontier 70
Belgium 67, 70
Bentley, E. C. 45
Berlin 107, 186
Berne 149
Beset by Secrecy and Beleaguered by Rivals 208
Between Silk and Cyanide 66, 208
Biberach 164, 192
Biography for Beginners 45
Bjorn 186
Blair, Lieutenant 14
Blanchain, Francis 164
Bletchley Park 145, 208
board games 90
Bordeaux 207, 208
Bowes, Christopher 79, 91
Boy's Own Paper 33
Breese, Flight Lieutenant J. C. 95, 106
Bremen 89, 120, 180
bridge 92
Briscoe, Professor H. V. A. 81
British Admiralty chart 54, 173
British Army 21, 25, 26, 82, 83, 119, 179
British Consulate, 100, 186
British Expeditionary Force (BEF) 25
British Intelligence in the Second World War 206
British Library 45, 79
British Military Attaché 128, 143

British Museum 33
Browns Sports Shop 106
Brunswick 109
Brunswick Printers 190, 191
Brussels, 205
Buckley, Lieutenant S. E. 156
Bulletin see MI9 *Bulletin*
Burma 72, 73
Bushy Park 83

Cairo 82
Calais 160
Calcutta 42, 82
Canada 123
Caspian Sea 81
Cavell, Edith Louisa 203, 205
Chamberlain, Neville 15
chess 89, 92, 93, 124, 125
China 37, 40, 72, 73
Christmas crackers 97
Churchill, Winston 8, 86, 93, 197, 206, 208
cigars 93
Civil War (American) 40
Clausthal 21
Clough, Brigadier A. B. 33
Coastal Defence Area 100
Cockleshell Heroes 207
Code Book, The 115
coded correspondence 115–141
coded letters 115–141, 143, 178
Col Powder lamination 94
Colditz *see* Oflag IVC (Colditz)
Colditz Story, The 160
Collier, Peter 198
Columbia 94
Comet Line 24, 148, 149

258 GREAT ESCAPES

Conduct of Work No. 48 15, 16
Connell, Charles 201
Constance, Lake 22, 149, 164, 189, 192
cribbage 89
cricket 89
Crockatt, Brigadier Norman R. 17, 18, 21, 24, 26, 30, 199, 202, 203, 205, 206, 207, 208, 209, 210, 213, 214
Crown and Anchor Mission 106
Cuba 40
Cumberland Pencil Company 98
Curragh 58

D-Day *see* Normandy Landings
Dansey, Colonel Claude 203, 205, 206, 208, 213
Danzig 46, 50, 54, 87, 173, 174, 175, 181, 185, 186, 187
darts 89
Defence Geographic Centre 60
Delhi 82
Denmark 64, 70, 73, 186
Derry, Sam 209
Dick 27
Dorington Committee 43
Double Eagle 46, 81, 193
draughts 89
Dunkirk 8, 13, 23, 85, 179, 206
Durnford, H. G. 27, 33
Dutch Girl 46, 60, 63

Embry, Air Marshall Sir Basil 201
Emerald 81
EMI 94

Empire Service League, The 106
ENIGMA 143, 144, 145
Escape Committee 109, 124, 125, 142, 143, 166, 181
escape packs 84, 102, 103, 104
escape-mindedness 9, 26, 169, 210
Escaping Club, The 22, 158
Europe 34, 52, 53, 59, 60, 70, 71, 73, 76, 100, 116, 149, 173, 187, 206
Europe Air series 49, 51, 59, 60, 61, 63, 81, 83, 192
Evans, Flight Lieutenant A. J. "Johnny" 13, 21, 22, 23, 25, 30, 33, 102, 119, 156, 158, 159
Evans, Philip Radcliffe 190

Far East 82
Ffrench-Mullen, Flight Lieutenant D. A. 95
Fight Another Day 24, 197, 210
Finland 73, 173
Finnish ships 186
Foot, M. R. D. 13, 18, 21, 112, 115, 116, 145, 162, 165, 171, 198, 199, 202, 203, 213
Foreign Office, 203
Forrest, Dr David 189
Fort 9 Ingolstadt 13, 22
Fortune, General Sir V. M. 127
France 24, 37, 63, 70, 82, 91, 100, 103, 117, 148, 206, 213
French Indo-China 73
French–Spanish border 67, 149
From Colditz in Code 116, 120

Gaddum, H. T. 82
Gaddum, P. W. 82
Garrison Map 33, 37, 40
Garrison Outline Map 40
Gdynia 85, 173, 175
General Map of Ireland 58
Geneva Convention, The 58, 98, 104, 115, 121
Geographia Ltd 34, 52
Geographic Section General Staff (GSGS) 83, 198, 199
Geographical Journal, The 7
GeoMapping Unit 193
George 27
George Philip & Son Ltd 79
Georgia, northwestern 10, 40
German Government Chart 173
German–Swiss border *see* Swiss Frontier
Germany 14, 17, 21, 23, 24, 27, 33, 46, 54, 64, 70, 73, 86, 91, 100, 102, 109, 113, 118, 125, 127, 128, 137, 139, 147, 148, 149, 159, 160, 163, 166, 167, 169, 171, 175, 178, 179, 180, 193, 210
Gestapo 143, 213
Gibraltar 205
Gironde, River 208
'goings' information 152
Gordon Highlanders 34
Government Communications Headquarters (GCHQ) 117
GR-107.94 31
gramophone records 89, 94, 97, 106, 107, 108, 125
Gray, Major 14

Great Escape, The 27, 30
Green, Julius 116, 117, 119, 120, 121, 122, 128, 129, 130, 145
Grouse 66
GSGS 4090 46, 63, 64
GSGS Series 3982 *see* Europe Air Series
Guérisse, Albert-Marie 149

Halex 103
Halifax, Viscount 15
Hamburg 181
Hamel, Lieutenant F. C. 125
Han Dynasty 37, 40
Hardanger Vidda 66
Harry 27
Harz Mountains 21
Heath, Wallis 190
Hidden Catch, The 201
Highgate Training School *see* Training School
Hinsley, F. H. 206
His Master's Voice (HMV) 94, 96
Hitler, Adolf 214
HMS *Erebos* 179
HMS *Hebe* 179
HMS *Hood* 179
HMS *Vindictive* 179
HMS *Warspite* 179
Holland *see* Netherlands
Holzminden prisoner of war camp 14, 27
Hooker, Mr 119
Horlicks 102
Hotine, Brigadier Martin 83
Houdini, Harry 21, 86
Hunan province China 37

Hunkin, Pilot Officer W. H. C. 95
Hutton, Christopher Clayton 9, 16, 20, 21, 33, 34, 37, 43, 52, 54, 63, 79, 81, 85, 86, 87, 91, 92, 94, 95, 96, 97, 99, 102, 104, 159, 197, 198, 199, 200, 201, 202, 211, 212

Imperial Chemical Industries (ICI) 81
Imperial College 81
Indian Office Library 45
Indonesia 72
Intelligence Corps Museum 100
International Map of the World 51, 70, 71, 73
Iran 82
Irish National Liberation Army 170
Italy 50, 54, 58, 63, 70, 82, 91, 125, 127, 143, 147, 209

James, Lieutenant D. P. 30, 109, 184, 185, 186, 187
Japan 72, 82
Java 72
Jig Saw Puzzle Club 106
John Bartholomew & Son Ltd 34, 35, 37, 46, 51, 52, 53, 54, 56, 58, 59, 75, 76, 79, 81, 193
John Waddington Ltd 45, 48, 63, 78, 79, 80, 81, 90, 91, 92, 193, 200
Joint Intelligence Committee 13, 15
Journey Has Been Arranged, A 201

Karachi 82
Keene, Thomas 208
Kennard, Captain Caspar 14
Kent County Cricket team 30
Kenya 81
Kenyon, Ley 27
Keswick 98
Kings Cross Station 81
Konigsberg 186
Korea 72
Korean War 165

Lake District (Keswick) 42, 43
Lancashire Penny Fund 97, 104
Langley, Lieutenant Colonel J. M. "Jimmy" 13, 18, 21, 23, 24, 30, 97, 116, 145, 149, 162, 171, 197, 198, 199, 202, 203, 205, 206, 210, 213
Laufen 179
Laupheim 164
League of Helpers 106
Lebanon 82
Leeds, University of 81
Lehman & Co., E. 93
Leipzig 163
Leisnig 163
Library of Congress 40
Licensed Victuallers' Sports Association 104, 124
Lille 23
Lisbon 205
Lizars 34
Lübeck 173, 175, 181, 182, 185, 186, 187
Ludo 88
Luteyn, Lieutenant Toni 24, 156, 162, 163, 165, 166, 167, 169

Madrid 149, 205
Maffey, Sir J. L. 58
Malay peninsula 72
manuscript escape maps 191, 192
Maps and Survey 34
Marks, Leo 66, 208
Marlag and Milag Nord 89, 109, 112, 115, 117, 119, 120, 121, 123, 125, 140, 160, 169, 178, 180, 181, 183, 184, 186
Marseilles 24, 149, 205, 206
Mary Queen of Scotts 116
McMullan, Professor David 117
Menzies, Sir Stewart 203, 205
Metropole Building (Metropole Hotel) 17, 25
MI1 13, 117
MI19 19
MI4 33, 34, 83, 199
MI5 13, 16
MI6 8, 15, 16, 149
MI9 *Bulletin* 19, 23, 26, 76, 77, 86, 87, 156, 159, 173, 174, 175, 178, 181, 182, 184
MI9 War Diary 15, 58, 78, 104, 118, 120, 127, 156, 169, 175, 181
MI9: Escape and Evasion 13, 18, 24
MI9a 19
MI9b 19
Middle East 34, 52, 53, 71, 82, 120
Military Cross 34
Military Intelligence Code and Cipher School 117
Military Intelligence, Deputy Director for 15

Military Intelligence, Directorate of 13, 15, 16
Military Operations and Intelligence, Directorate of 13
Military Operations, Directorate of 13, 58
Military Survey, Directorate of (D.Survey) 43, 45, 49, 52, 60, 63, 70, 71, 78, 83, 198, 199, 211
Ming Dynasty 40
Ministry of Defence 8, 45
Ministry of Supply 63, 78, 80, 81, 82, 90, 91, 92, 104
MIS-X 92, 120
Miscellaneous Maps 75
Monopoly 88, 90, 91, 198
Morning Post, The 8
Mozambique 8
Munich 108, 179
Mysore 82

National Archives 8, 45
National Library of Scotland 58
Neave, A. M. S. "Airey" 24, 25, 30, 96, 107, 108, 116, 147, 156, 160–171, 189, 191, 203, 205, 210, 213
Netherlands 14, 67, 70
Netherlands East Indies Army, 162
Normandy Landings 70, 178, 179, 183, 210, 211, 212
Norsk Hydro facility 66
Northamptonshire Regiment, 156
Norway 46, 63, 64, 75, 76, 87, 91
Nouveau, Louis 164
Nuremburg 22, 158,

O'Leary, Patrick Albert 149
O'Sullivan, Lieutenant Commander M. J. A. 178
Official Secret 21, 52, 85, 199, 201, 202
Official Secrets Act 201
Oflag IVC (Colditz) 22, 24, 25, 95, 96, 97, 106, 107, 109, 113, 120, 156, 160, 160–171, 189
Oflag IX 128
Oflag VB (Biberach) 164, 191
Oflag VIB 128
Oflag VIIC/H 179
Oflag VIIC/Z (Titmoning) 127, 180
Oflag XI (Braunschweig) 190, 191
Old Ladies Knitting Committee, The 106
Operation Frankton 207
Operation Market Garden 63
Operation Smash-Hit 97
Ordnance Survey 42, 43, 79
Oslo 64
Overlord 116

Park Saw Mills 21
Part of Northern Georgia 10
Pat Line 24, 148
Pegram, Lieutenant Colonel J. C. 33, 42
Pencil Museum 98
pencils 98, 99
Penck, Albert 70
Per Ardua Libertas 16, 92, 93, 99, 105
Pesant, Roberto 167, 169
Petit Journal, Le 205

INDEX 261

Pfullendorf 164
Philips *see* George Philip & Son Ltd
Philpot, Flight Lieutenant Oliver 173, 186, 187
playing cards 91, 92, 93
Popular Science 202
Portugal 67, 70
Prassinos, Mario 164
Prisoner's Progress, A 30, 184, 185
Prisoners' Leisure Hours Fund 104, 105, 106, 165
Pryor, Lieutenant John 27, 106, 109, 115, 116, 117, 120, 122, 123, 124, 125, 127, 128, 130, 131, 133, 134, 135, 136, 139, 140, 141, 147, 160, 178, 179, 180, 181, 183, 184, 186, 187
purse maps 100
purses 99
Pyrenees 70, 149

Quartermaster's department 102
Queen Elizabeth I 116

RAF Benson 100
Ramsen 164, 166, 167
Red Cross 98, 104, 108, 122
Red Cross parcels 104, 107, 108, 180, 191
Regensburg 163
Reid, Captain P. R. 96, 160, 166, 167
renten 137
Republic of Ireland 58
Resistance 149

Rhodes, Lieutenant Commander 119
Rjukan 64
Rome 82
Royal Air Force 21, 25, 26, 58, 64, 83, 87, 99, 100, 103, 117, 118, 119, 122, 185, 186, 206
Royal Air Force Escaping Society 209
Royal Air Force Museum 93, 95
Royal Army Signals Corps 166
Royal Artillery 24, 160
Royal Corps of Engineers 43, 190
Royal Flying Corps 14, 21, 156
Royal Geographic Society 70, 71, 197, 198
Royal Marines 26, 207
Royal Navy 21, 26, 83, 117, 179, 183
Royal Navy Volunteer Reserve 125, 184
Royal Scots Guards 17
Ruffec 208
Rugby School 33
Russia 53, 73

Sagan north 186
Salz, River 179
Sandbostel 89, 120, 121, 139, 180
Saturday 24, 116
Saturday at MI9 25, 205
Scandia 185
Scandinavia 53, 59, 73
Schaffhausen *see* Schaffhausen Salient

Schaffhausen Salient 22, 23, 45, 54, 76, 86, 95, 147, 149, 152, 156, 158, 159, 160, 163, 164, 173, 192
School of Geography, Plymouth University 193
School of Mathematics and Physics, Plymouth University 117
Schwachenreuter 164
Secret Intelligence Service (SIS) 8, 15, 24, 149, 203, 206, 208, 213
Section D 19, 25
Section W 19
Section X 19, 63, 64
Section Y 19, 118
Section Z 19, 20, 214
[Series 43] 51, 67, 71, 73, 175, 211, 212
[Series 44] 52, 70, 71, 72, 73
[Series FGS] 52, 73, 75
Shamrock 58
Sheet 19D 66
Sheet 26B 64
Sheet 26D 64
Sheet 43A 67, 70, 149
Sheet 43B 45, 67, 70
Sheet 43C 70, 212
Sheet 43D 48, 70
Sheet 43K East, 70
Sheet 43K West, 70
Sheet 44G 71
Sheet 44H 72
Sheet 44N 73
Sheet 44O 73
Sheet 9Ca 100
Sheet 9J3 76
Sheet A 46, 54, 55, 193

Sheet A1 86, 156
Sheet A10 85
Sheet A11 85
Sheet A2 (Bulletin) 86, 156, 165, 167, 170
Sheet A3 50, 173, 175
Sheet A4 50, 54, 173, 174, 186
Sheet A6 54, 189
Sheet C 49, 50, 54
Sheet D (FGS) 48, 75
Sheet E35 West 64, 66
Sheet F 52
Sheet H2 54
Sheet J3 50, 76
Sheet J4 50
Sheet J5 54, 58
Sheet J6 54
Sheet J7 54
Sheet J8 54, 58
Sheet K3 53, 54, 59
Sheet K4 59
Sheet L32-2/Konstanz 192
Sheet Q 81
Sheet S2 51
Sheet S3 51
Sheet T2/T4 81
Sheet Y 23, 26, 95, 147, 152, 156, 159, 165, 167, 170
Shelmerdine, Flight Lieutenant J. H. 100
Sherman, General W. T. 40
Sinclair, Admiral Quex 203
Singen 164, 167
Singh, Simon 115, 116
Smith-Cumming, Sir George Mansfield 203
Snakes and Ladders 88
Södertälje 186

Somme 21, 156
Spain 24, 67, 70, 100, 148, 149, 164, 213
Sparks, Bill 207
Special Air Service (SAS) 213
Special Forces Club 209
Special Operations Executive (SOE) 63, 64, 66, 120, 198, 200, 206, 207, 209, 213
Special Watch List 126, 169
squash racquets 89
Stalag Luft III (Sagan) 27, 30, 95, 106, 107, 108, 123, 128, 132, 142, 186, 187, 189
Stalag VIIA (Moosburg) 108
Stalag XXI D (Posen) 95
Stettin 173, 175, 178, 184
Stockholm 128, 143, 149, 186
Stuart, Sir Campbell 13, 15
Survey 2 45
Survey Division of Headquarters Strategic Allied Command (SAC) 71, 72
Survey of India 42
Sweden 9, 64, 91, 148, 173, 178, 185, 187
Swedish border 64
Swedish ships 50, 173, 174, 178, 186
Swiss Frontier *see* Swiss–German border
Swiss–German border 22, 23, 45, 59, 67, 70, 123, 125, 149, 152, 156, 158, 159, 160, 163, 164, 166, 167, 169, 171, 181, 191, 192, 193
Switzerland 9, 24, 63, 70, 147, 148, 149, 152, 158, 160, 163, 164, 169, 170, 171, 191

Tehran, 82
Templar, Field-Marshal Sir Gerald 13
Territorial Army 117, 160
Thailand 73
They Have Their Exits 25, 160, 162, 167, 169, 189
Third Reich 213
Thomas Salter 93
Thomas, Colonel R. H. 42
tobacco pouches 100
Tom 27
Toscanini 97
Town, Henry 81
Training School 9, 18, 19, 21, 25–26, 86, 118, 159, 175
Trinity Hall 23
Turing, Alan 145
Turkey, 53, 67, 82
Turner, Squadron Leader W. H. N. 95

Ulm 163, 164
uniforms 99, 100, 102
United States of America 10, 40, 82, 92, 201, 214
Uppingham 23
US War Department 33

van Lier, Peggy 24
Vermok 66

Waddington's *see* John Waddington Ltd
Wagner 97
Walsingham, Sir Francis 116
War Cabinet 205, 206

War Office 21, 34, 42, 43, 45, 46, 70, 83, 109, 118, 166, 198, 201, 202, 203
Wardle, Flight Lieutenant H. N. 166
Washington DC 40
Watson, Norman 78, 92
Watson, Victor H. 78, 80, 91, 200
Wechselmünde 173
Wells, Lieutenant John 181, 183, 186
Western Front 21
Westertimke 139, 183
Whiston, Eric 82
Whitehall 83
Wiggins, Lieutenant Colonel W. D. C. 52, 71, 72
Wills, W. D. & H. O. 102
Wilton Park 18
Winchester 22
Winterbotham, Major 128
Wolverhampton 23
Wooler, John 94, 95
Woollatt, Hugh 164, 191
World War I 17, 21, 22, 27, 34, 102, 156, 159, 197, 199, 203, 205
World War II 9, 22, 31, 34, 117, 149, 192, 197, 198, 213, 214
Worsley, John 89, 181
Wülzburg, Castle of 126

Yeomanry, the Yorkshire Regiment 21
Ypres 34, 35, 156

Zones of France map 37